Collins

COMPLETE PLUMBING & CENTRAL HEATING

Changes to electrical regulations:
From 1 January 2005 electrical installation work in England and Wales must meet the requirements of a new section of the Building Regulations, Part P. Similar regulations apply in Scotland and Northern Ireland. There is a legal requirement for you to notify your local Building Control department regarding all major electrical installations including work in your kitchen, bathroom, utility room and outdoors, and to have such work carried out by a competent electrician. However, you can still do certain minor repairs, replacements and maintenance work, such as replacing socket outlets, switches and ceiling roses, or make other alterations to an existing circuit except in bathrooms, kitchens, utility rooms or outdoors. For full details, and especially if you are in any doubt about what you are able to do yourself, contact your local authority's Building Control department.

See also:
www.partp.co.uk
www.odpm.gov.uk/electricalsafety
www.scotland.gov.uk/build_regs
www2.dfpni.gov.uk/buildingregulations

Collins

COMPLETE PLUMBING
& CENTRAL HEATING

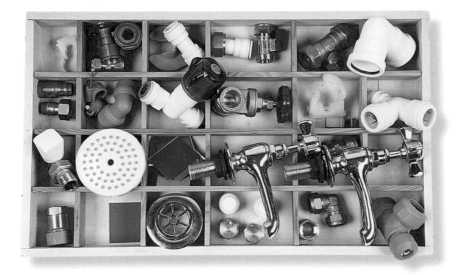

fix it yourself and save money -
from changing a washer to installing a shower

Published by HarperCollins*Publishers*, London

This edition first published in 2003. Most of the text and illustrations in this book previously appeared in the *Collins Complete DIY Manual*, published in 2001.

ISBN-13 978 0 00 716441 7
ISBN-10 0 00 716441 6

Copyright © 1988, 1995, 1999
HarperCollins*Publishers*

The material for this book was created exclusively for HarperCollins*Publishers* by Jackson Day Jennings Ltd trading as Inklink

Authors Albert Jackson & David Day

Editorial director Albert Jackson

Design, art direction and project management Simon Jennings

Text editor Peter Leek

Design and production assistant Amanda Allchin

Consultants Roger Bisby & John Dees

Illustrations editor David Day

Illustrators Robin Harris & David Day

Additional illustrations Brian Craker, Michael Parr & Brian Sayers

Photographer Ben Jennings

Additional photography Neil Waving and Paul Chave

Proofreader Mary Morton

Research editor and technical consultant Simon Gilham

The CIP catalogue record for this book is available from the British Library

Colour origination by Colourscan, Singapore

Printed and bound by CPI Bath

PLEASE NOTE
Great care has been taken to ensure that the information contained in this book is accurate. However, the law concerning Building Regulations, planning, local bylaws and related matters is neither static nor simple. A book of this nature cannot replace specialist advice in appropriate cases and therefore no responsibility can be accepted by the publisher or by the authors for any loss or damage caused by reliance upon the accuracy of such information.

CONTENTS

Cross-references
Since there are few DIY projects that do not require a combination of skills, you might have to refer to more than one section of the book. The list of cross-references at the bottom of each page will help you locate relevant sections or specific information related to the job in hand.

Plumbing systems

The unprecedented supply of tools and easy-to-use hardware has encouraged DIY enthusiasts to tackle their own plumbing repairs and improvements. Almost every aspect is now catered for – with a wide range of metal and plastic pipework and attractive fittings and appliances, both for new installations and for refurbishments.

The advantages of DIY plumbing

Having the wherewithal to tackle your own plumbing installations and repairs can save you the cost of hiring professionals – and that can amount to a substantial sum of money. It also avoids the distress and inconvenience of ruined decorations, and the expense of replacing rotted household timbers where a slow leak has gone undetected. Then there's the saving in water. A dripping tap wastes gallons of water a day – and if it's hot water, there's the additional expense of heating it. A little of your time and a few pence spent on a washer can save you pounds.

Water systems

Generally, domestic plumbing incorporates two systems. One is the supply of fresh water from the 'mains', and the other is the waste or drainage system that disposes of dirty water. Both of the systems can be installed in different ways (see opposite).

a temporary mains failure; the major part of the supply is under relatively low pressure, so the system is reasonably quiet; and because there are fewer mains outlets, there is less likelihood of impure water being siphoned back into the mains supply.

Stored-water system (Indirect)
The majority of homes are plumbed with a stored-water supply system. The storage tank in the loft and the cold-water tap in the kitchen are fed directly from the mains; so possibly are your washing machine, electric shower(s) and outside tap. But water for baths, washbasins, flushing WCs and some types of shower is drawn from the storage tank, which should be covered with a purpose-made lid to protect the water from contamination. Drinking water should only be taken from the cold-water tap in the kitchen.

Cold water from the storage tank is fed to a hot-water cylinder, where it is heated by a boiler, indirectly, or by an immersion heater to supply the hot taps. The water pressure at the various taps in the house depends on the height (or 'drop') from the tank to the tap.

A stored-water system provides several advantages. There is adequate water to flush sanitaryware during

Mains-fed system (Direct)
Many properties now take all their water directly from the mains – all the taps are under high pressure, and all of them provide water that's suitable for drinking. This development has come about as a result of limited loft space that precludes a storage tank and the introduction of non-return check valves, which prevent drinking water being contaminated. Hot water is supplied by a combination boiler or a multipoint heater; these instantaneous heaters are unable to maintain a constant flow of hot water if too many taps are running at once. Some systems incorporate an unvented cylinder, which stores hot water but is fed from the mains.

A mains-fed system is cheaper to install than an indirect one. Another advantage is mains pressure at all taps; and you can drink from any cold tap in the house. With a mains-fed system there's no plumbing in the loft to freeze.

Drainage

Waste water is drained in one of two ways. In houses built before the late 1950s, water is drained from baths, sinks and basins into a waste pipe that feeds into a trapped gully at ground level. Toilet waste feeds separately into a large-diameter vertical soil pipe that runs directly to the underground main drainage network.

With a single-stack waste system, which is installed in later buildings, all waste water drains into a single soil pipe – the one possible exception being the kitchen sink, which may drain into a gully.

Rainwater usually feeds into a separate drain, so that the house's drainage system will not be flooded in the event of a storm.

Water bylaws govern the way you can connect your plumbing system to the public water supply. These laws are intended to prevent the misuse, waste and contamination of water. Your local water supplier will provide you with the relevant information about inspection requirements and possible certification for new work and for major alterations.

Before undertaking work

The Building Regulations on drainage are designed to protect health and safety. Before undertaking work on your soil and waste pipes or drains (except for emergency unblocking) you need to contact the building-control department of your local authority.

You are required to give five days notice to your local water supplier before altering or installing a lavatory cistern, bidet, shower pump, hosepipe supply, or any installation, such as a garden tap or shower, that could cause dirty water to be siphoned back into the supply of drinking water.

● **Wiring Regulations**
When making repairs or improvements to your plumbing, make sure you don't contravene the electrical Wiring Regulations. All metal plumbing has to be bonded to earth. If you replace a section of metal plumbing with plastic, it is important to reinstate the earth link. (See far right).

Reinstate the link
If you replace a section of metal plumbing with plastic, you may break the path to earth – so make sure you reinstate the link. Bridge a plastic joint in a metal pipe with an earth wire and two clamps. If you are in any doubt, consult a qualified electrician.

MAINS-FED SYSTEM (opposite)

1 Water-supplier's stopcock
May include water meter.

2 Service pipe

3 Main stopcock

4 Rising main
Supplies water directly to cold-water taps and WCs etc.

5 Water heater or combination boiler

6 Unvented storage cylinder
(Not required for instantaneous heaters)

7 Single-stack soil pipe
WC, handbasin, bath and shower drain into the stack. The stack may be fitted with an air-admittance valve terminating inside the house.

8 Sink waste
Water from the sink drains into a trapped gully.

9 Trapped gully

☞ **SEE ALSO:** Draining rainwater 7, Garden tap 48, Earthing 68–70, Supplementary bonding 69–70, 81

Direct and indirect systems

Stored-water system
● **Central heating** omitted for clarity.

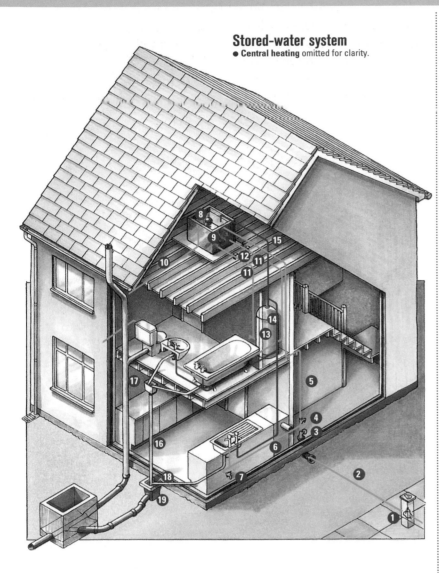

STORED-WATER SYSTEM

1 Water company stopcock
The water company uses this stopcock to turn off the supply to the house. Make sure it can be located quickly in an emergency.

2 Service pipe
From the water company stopcock onwards, the plumbing becomes the responsibility of the householder.

3 Household stopcock
The water supply to the house itself is shut off at this point.

4 Draincock
A draincock here allows you to drain water from the rising main.

5 Rising main
Mains-pressure water passes to the cold-water storage tank via the rising main.

6 Drinking water
Drinking water is drawn off the rising main to the kitchen sink.

7 Garden tap
The water company allows a garden tap to be supplied with mains pressure, provided it is fitted with a check valve.

8 Float valve
This valve shuts off the supply from the rising main when the cistern is full.

9 Cold-water storage tank
Stores from 230 to 360 litres (50 to 80 gallons) of water. Positioned in the roof, the tank provides sufficient 'head', or pressure, to feed the whole house.

10 Overflow pipe
Also known as a warning pipe, it prevents an overflow by draining water to the outside.

11 Cold-feed pipes
Water is drawn off to the bathroom and to the hot-water cylinder from the storage tank.

12 Cold-feed valves
Valves at these points allow you to drain the cold water in the feed pipe without having to drain the whole tank as well.

13 Hot-water cylinder
Water is heated and stored in this cylinder.

14 Hot-feed pipe
All hot water is fed from this point.

15 Vent pipe
Allows for expansion of heated water and enables air to be vented from the system.

16 Waste pipe
Surmounted by a hopper head, it collects water from basin and bath.

17 Soil pipe
Separate pipe takes toilet waste to main drains.

18 Kitchen waste pipe
Kitchen sink drains into same gully as waste pipe from upstairs.

19 Trapped gully

● **Water meters**
Instead of paying a flat-rate water charge based upon the size of your home, you can opt to have your water consumption metered so you pay for what you use. For two people living in a large house, the savings can be considerable. Water meters are fitted to the incoming mains, usually outside at the supplier's stopcock, where they can be read more easily.

Mains-fed system
● **Central heating** omitted for clarity.

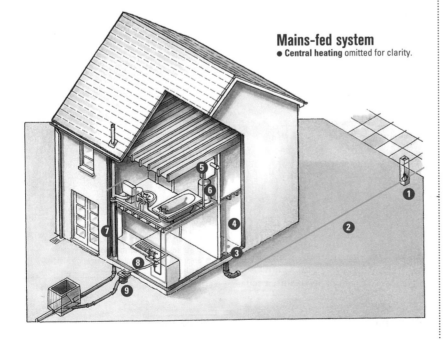

☞ **SEE ALSO: Wet central heating 53**

Draining the system

You will have to drain at least part of any plumbing system before you can work on it; and if you detect a leak, you will have to drain the relevant section quickly. So find out where the valves, stopcock and draincocks are situated, before you're faced with an emergency.

Draining cold-water taps and pipes

● Turn off the main stopcock on the rising main to cut off the supply to the kitchen tap (and to all the other cold taps on a direct system).
● Open the tap until the flow ceases.
● To isolate the bathroom taps, close the valve on the appropriate cold-feed pipe from the storage tank and open all taps

on that section. If you can't find a valve, rest a wooden batten across the tank and tie the arm of the float valve to it. This will shut off the supply to the tank, so you can empty it by running all the cold taps in the bathroom. If you can't get into the loft, turn off the main stopcock, then run the cold taps.

Draining hot-water taps and pipes

● Turn off immersion heater or boiler.
● Close the valve on the cold-feed pipe to the cylinder and run the hot taps. Even when the water stops flowing, the cylinder will still be full.
● If there's no valve on the cold-feed pipe, tie up the float-valve arm, then turn on the cold taps in the bathroom

to empty the storage tank. (If you run the hot taps first, the water stored in the tank will flush out all your hot water from the cylinder.) When the cold taps run dry, open the hot taps. In an emergency, run the hot and cold taps together in order to clear the pipes as quickly as possible.

Draining a WC cistern

● To merely empty the WC cistern itself, tie up its float-valve arm and flush the WC.
● To empty the pipe that supplies the cistern, either turn off the main stopcock on a direct system or, on an

indirect system, close the valve on the cold feed from the storage tank. Alternatively, shut off the supply to the storage tank and empty it through the cold taps. Flush the WC until no more water enters its cistern.

Draining the cold-water storage tank

● To drain the storage tank in the roof space, close the main stopcock on the rising main, then open all the cold taps

in the bathroom (hot taps on a direct system.) Bail out the residue of water at the bottom of the tank.

Draining the hot-water cylinder

● If the hot-water cylinder springs a leak (or you wish to replace it), first turn off the immersion heater and boiler, then shut off the cold feed to the cylinder from the storage tank (or drain the cold-water storage tank – see above). Run hot water from the taps.
● Locate a draincock from which you can drain the water remaining in the cylinder. It is probably located near the base of the cylinder, where the cold feed from the storage tank enters. Attach a hose and run it to a drain or sink that is lower than the cylinder. Turn the square-headed spindle on the draincock till you hear water flowing.
● Water can't be drained if the washer

is baked onto the draincock seating, so disconnect the vent pipe and insert a hosepipe to siphon the cylinder.
● Should you want to replace the hot-water cylinder, don't disconnect all its pipework until you have drained the cylinder completely. If the water is heated indirectly by a heat-exchanger, there will be a coil of pipework inside the hot-water cylinder that is still full of water. This coil can be drained via the stopcock on the boiler after you have shut off the mains supply to the small feed-and-expansion tank, which is located in the roof space. Switch off the electrical supply to the central-heating system.

Saving hot water
If your gate valve won't close off and you don't want to drain all the hot water, you can siphon the water out of the cold tank with a garden hosepipe. While the tank is empty, replace the old gate valve.

● **Sealed central-heating systems**
A sealed system (see SEALED CENTRAL-HEATING SYSTEMS) does not have a feed-and-expansion tank – the radiators are filled from the mains via a flexible hose known as a filling loop. The indirect coil in the hot-water cylinder is drained as described right, though you might have to open a vent pipe that is fitted to the cylinder before the water will flow.

Unless you divide up the system into relatively short pipe runs with valves, you will have to drain off a substantial part of a typical plumbing installation even for a simple washer replacement.

● Install a gate valve on both the cold feed pipes running from the cold-water storage tank. This will eliminate the necessity for draining off gallons of water in order to isolate pipes and appliances on the low-pressure cold- and hot-water supply.
● When you are fitting new taps and appliances, take the opportunity to fit miniature valves on the supply pipes. In future, when you have to repair an individual tap or appliance, you will be able to isolate it in moments.

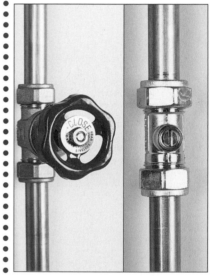

Gate valve
Fit a gate valve to the cold-feed pipes from the storage tank.

Miniature valve
Fit a miniature valve to the supply pipes below a sink or basin.

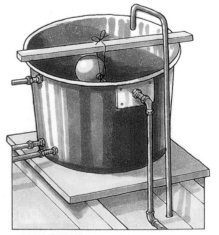

Closing a float valve
Cut off the supply of water to a storage tank by tying the float arm to a batten.

☛ **SEE ALSO:** Cylinders 50-1, Radiators 55, Consumer unit 68

DRAINING AND REFILLING THE SYSTEM

Partially drain the plumbing system if you intend to leave the house unoccupied for a few days during winter – if possible, leave the central heating on a low setting. For longer periods of absence at any time of the year, you may want to take the precaution of draining the system completely.

Partial drain-down
● Add special antifreeze to the central-heating feed-and-expansion tank and set the heating to come on for a short period twice a day.
● Turn off the main stopcock.
● Open all the taps to drain the house's water system.

Attach hosepipe to draincock

Full drain-down
● Switch off and extinguish the water heater and/or boiler.
● Turn off the main stopcock and, if possible, the water company stopcock outside.
● Open all taps in the house to drain the pipework.
● Open the draincock at the base of the hot-water cylinder. If there are draincocks in the rising main and in any other low pipework, drain the water from these too.
● Flush the WCs.
● Drain the boiler and radiator circuits at the lowest points on the pipe runs.
● Add salt to the WC pan to prevent the trap water freezing.

Refilling the system
● Close all taps and draincocks.
● Turn on the main stopcock.
● Turn on taps and allow water and air to escape. As the system fills, check that float valves are operating smoothly.

Curing an airlock
Air trapped in the system can cause a tap to splutter. The answer is to force the air out by using mains pressure.
    Attach a length of hose between the affected tap and any mains-fed cold-water tap. Leave both taps open for a short while, and then try the airlocked tap again. Repeat if necessary, until the water runs freely.

Emergency repairs

It pays to master the simple techniques for coping with emergency repairs – in order to avoid the inevitable damage to your home and property, as well as the high cost of calling out a plumber at short notice. All you need is a simple tool kit and a few spare parts.

Thawing frozen pipes

If water won't flow from a tap during cold weather, or a tank refuses to fill, a plug of ice may have formed in one of the supply pipes. The plug cannot be in a pipe supplying taps or float valves that are working normally, so you should be able to trace the blockage quickly. In fact, freezing usually occurs first in the roof space.
    As copper pipework transmits heat quickly, use a hairdryer to gently warm the suspect pipe, starting as close as possible to the affected tap or valve and working along it. Leave the tap open, so water can flow normally as soon as the ice thaws. If you can't heat the pipe with a hairdryer, wrap it in a hot towel or hang a hot-water bottle over it.

Preventative measures
Insulate pipework and fittings to stop them freezing, particularly those in the loft or under the floor. If you're going to leave the house unheated for a long time during the winter, drain the system (see left). Cure any dripping taps, so leaking water doesn't freeze in your drainage system overnight.

Dealing with a punctured pipe

Unless you are absolutely sure where your pipes run, it is all too easy to nail through one of them when fixing a loose floorboard. You may be able to detect a hissing sound as water escapes under pressure, but more than likely you won't notice your mistake until a wet patch appears on the ceiling below, or some problem associated with damp occurs at a later date. While the nail is in place, water will leak relatively slowly, so don't pull it out until you have drained the pipework and can repair the leak. If you pull out the nail by lifting a floorboard, replace the nail immediately.
    If you plan to lay fitted carpet, you can paint pipe runs on the floorboards to avoid such accidents in future.

Patching a leak

During freezing conditions, water within a pipe turns to ice, which expands until it eventually splits the walls of the pipe or forces a joint apart. Copper pipework is more likely to split than lead, which can stretch to accommodate the expansion and thus survive a few hard winters before reaching breaking point. Temporarily patch copper or lead pipes as described right – but close up a split in lead beforehand by tapping the pipe gently with a hammer. Arrange to replace the old lead with copper pipe as soon as you have contained the leak.
    The only other reason for leaking plumbing is mechanical failure – either through deterioration of the materials or because a joint has failed and is no longer completely waterproof.
    If possible, make a permanent repair, by inserting a new section of pipe or replacing a leaking joint. (If it is a compression joint that has failed, try tightening it first.) For the time being, however, you may have to make an emergency repair. Drain the pipe first unless it is frozen, in which case make the repair before it thaws.

Binding a leaking pipe
For a temporary repair, cut a length of garden hose to cover the leak and slit it lengthwise, so you can slip it over the pipe. Bind the hose with two or three hose clips; or, using pliers, twist wire loops around the hose.
    Alternatively, use amalgamating tape made for binding damaged pipes.

Patching with epoxy putty
Epoxy putty adheres to most metals and hard plastic and will produce a fairly long-term repair, although it is better to insert a new length of pipe. The putty is supplied in two parts which begin to harden as soon as they are mixed together, giving about 20 minutes to complete the repair.
    First clean a 25 to 50mm (1 to 2in) length of pipe on each side of the leak, using wire wool. Mix the putty and press it into the hole or around a joint, building it to a thickness of 3 to 6mm ($\frac{1}{8}$ to $\frac{1}{4}$in). It will cure to full strength within 24 hours, but you can run low-pressure water immediately if you bind the putty with self-adhesive tape.

Thawing a frozen pipe
Play a hairdryer gently along a frozen pipe, working away from the blocked tap or valve.

Closing a split pipe
In an emergency, close a split by tapping the pipe with a hammer before you bind it. This works particularly well with lead pipe.

Binding a split pipe
Bind a length of hosepipe around a damaged pipe, using hose clips or wire. Alternatively, use an amalgamating tape.

Smoothing epoxy putty
When patching a hole with epoxy putty, smooth it with a damp soapy cloth to give a neat finish.

☞ **SEE ALSO:** Compression joints 20, Joining pipes 20–4

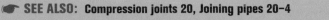

Repairing a leaking tap

A tap may leak for a number of reasons – none of them difficult to deal with. When water drips from a spout, for example, it is usually the result of a faulty washer; and if the tap is old, the seat against which the washer is compressed may be worn, too. If water leaks from beneath the head of the tap when it's in use, the gland packing or O-ring needs replacing. When you are working on a tap, insert the plug and lay a towel in the bottom of the washbasin, bath or sink to catch small objects.

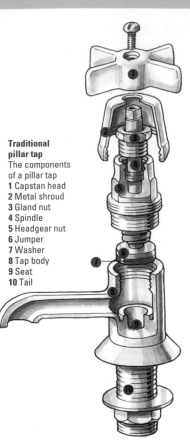

Traditional pillar tap
The components of a pillar tap
1 Capstan head
2 Metal shroud
3 Gland nut
4 Spindle
5 Headgear nut
6 Jumper
7 Washer
8 Tap body
9 Seat
10 Tail

Replacing a washer

To replace the washer in a traditional bib or pillar tap, first drain the supply pipe, then open the valve as far as possible before you begin dismantling either kind of tap.

If the tap is shrouded with a metal cover, unscrew it by hand or use a wrench, taping the jaws to protect the chrome finish.

Lift up the cover to reveal the headgear nut just above the body of the tap. Slip a narrow spanner onto the nut and unscrew it (**1**) until you can lift out the entire headgear assembly.

The jumper to which the washer is fixed fits into the bottom of the head-gear. In some taps the jumper is removed along with the headgear (**2**), but in other types it will be lying inside the tap body.

The washer itself may be pressed over a small button in the centre of the jumper (**3**) – in which case, prise it off with a screwdriver. If the washer is held in place by a nut, it can be difficult to remove. Allow penetrating oil to soften any corrosion; then, holding the jumper stem with pliers, unscrew the nut with a snug-fitting spanner (**4**). (If the nut won't budge, replace the whole jumper and washer.)

Fit a new washer and retaining nut, then reassemble the tap.

Removing a shrouded head from a tap
On most modern taps the head and cover is in one piece. You will have to remove it to expose the headgear nut. Often a retaining screw is hidden beneath the coloured hot/cold disc in the centre of the head. Prise out the disc with the point of a knife (**1**). If there's no retaining screw, simply pull the head off (**2**).

1 Prise out the disc

2 Pull the head off

1 Loosen headgear nut

2 Lift out headgear

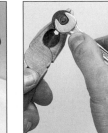

3 Prise off washer

4 Or undo fixing nut

Curing a dripping ceramic-disc tap

In theory ceramic-disc taps are maintenance free, but faults can still occur. Since there's no washer to replace, you have to replace the whole inner cartridge when the tap leaks. However, before you proceed, check that the lower seal is not damaged, as this can cause the tap to drip.

Turn off the water and remove the headgear from the tap body by turning it anticlockwise with a spanner (**1**).

Remove the cartridge and examine it for wear or damage (**2**). Cleaning any debris off the ceramic discs might be all that is required; but if a disc is cracked, then you will need a new cartridge. Cartridges are handed – left (hot) and right (cold) – so be sure to order the correct one.

At the same time, examine the rubber seal on the bottom of the cartridge. If this is worn or damaged, it will cause the tap to drip. If need be, replace the seal with a new one (**3**).

SERVICING REVERSE-PRESSURE TAPS

You can replace the washer in a reverse-pressure tap without turning off the water supply. Loosen the locking nut with a spanner (**1**); it has a left-hand thread, so you need to turn it clockwise (when viewed from above).

To release the tap body into your hand (**2**), turn the tap on – the initial jet of water will stop automatically. Gently tap the body on a wooden surface to eject the finned nozzle from inside. Prise off the combined jumper and washer, and replace it (**3**).

Reverse-pressure tap

1 Loosen locking nut

1 Unscrew the cartridge

2 Lift out and examine

3 Replace a worn rubber seal

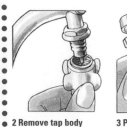

2 Remove tap body

3 Prise off jumper

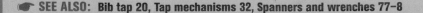

☞ **SEE ALSO:** Bib tap 20, Tap mechanisms 32, Spanners and wrenches 77–8

Repairing seats and glands : REPLACING O-RINGS

Regrinding the seat

If a tap continues to drip after you have replaced the washer, the seat is probably worn, allowing water to leak past the washer. One way to cure this is to grind the seat flat with a special reseating tool available from plumber's suppliers.

Remove the headgear and jumper, so you can screw the reseating tool into the body of the tap. Adjust the cutter until it is in contact with the seat, then turn the handle to smooth the metal **(1)**.

Alternatively, you can cover the old seat with a nylon liner that is sold with a matching jumper and washer **(2)**. Drop the liner over the old seat, replace the jumper and assemble the tap. Finally, close the tap to force the liner into position.

On a mixer tap each valve is usually fitted with a washer, as on conventional taps, but in most mixers the gland packing (see left) has been replaced by a rubber O-ring.

Having removed the shrouded head, take out the circlip holding the spindle in place **(1)**. Remove the spindle and slip the O-ring out of its groove **(2)**. Replace the old ring with a new one, using silicone grease as a lubricant, then reassemble the tap.

GLAND PACKING

Gland packing
Older-style taps are sealed with water-tight packing around the spindle.

1 Revolve the tool to smooth the seat

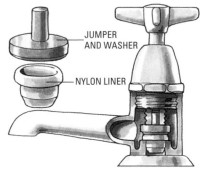

JUMPER AND WASHER

NYLON LINER

2 Repair a worn seat with a nylon liner

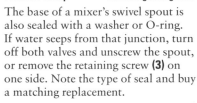

1 Remove circlip　　**2 Roll ring from groove**

The base of a mixer's swivel spout is also sealed with a washer or O-ring. If water seeps from that junction, turn off both valves and unscrew the spout, or remove the retaining screw **(3)** on one side. Note the type of seal and buy a matching replacement.

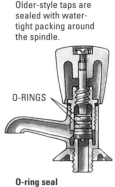

O-RINGS

O-ring seal
Modern taps are sealed with rubber rings, in place of gland packing.

Curing a leaking gland

The head of a tap is fixed to a shaft or spindle, which is screwed up or down to control the flow of water. The spindle passes through a gland – also known as a stuffing box – on top of the headgear assembly. A watertight packing is forced into the gland by a nut to prevent water leaking past the spindle when the tap is turned on. If water drips from under the head of the tap, the gland packing has failed and needs replacing.

Some taps incorporate a rubber O-ring that slips over the spindle to perform the same function as the packing (see right).

Replacing the gland packing
There is no need to turn off the supply of water to replace gland packing: just make sure the tap is turned off fully.

To remove a cross or capstan head, expose a fixing screw by picking out the plastic plug in the centre of the head, or look for a screw holding it at the side. Lift off the head by rocking it from side to side, or tap it gently from below with a hammer.

If the head is stuck firmly, open the tap as far as possible, unscrew the cover, and wedge wooden packing between it and the headgear **(1)**. Closing the tap will then jack the head off the spindle.

Once you have removed the head and cover, try to seal the leak by tightening the gland nut. If that fails, remove the nut and pick out the old packing with a small screwdriver.

To replace the packing, either use the special fibre string available from plumbers' merchants or twist a thread from PTFE (polytetrafluorethylene) tape. Wind the string around the spindle, and pack it into the gland with the screwdriver **(2)**.

3 Remove the screw to release the mixer spout.
You can use a cranked screwdriver (below) if the retaining screw is located behind the swivel spout.

Stopcocks and valves

Stopcocks and gate valves are used so rarely that they often fail to work just when they are needed.

Make sure that they are operating smoothly by closing and opening them from time to time. If their spindles move stiffly, lubricate them with a little penetrating oil. A stopcock is fitted with a standard washer, but as it is hardly ever under pressure it is unlikely to wear. However, the gland packing (see left) on both stopcocks and gate valves may need attention.

1 Jack the head off a tap with wooden packing

2 Stuff string or a thread of PTFE tape into the gland

Maintaining cisterns and storage tanks

The mechanisms used in WC cisterns and storage tanks are probably the most overworked of all plumbing components, so servicing is required from time to time to keep them operating properly. You can get the spare parts you need from plumbers' merchants and DIY stores.

Low-level WC cisterns are particularly easy to service, but even an old-style wall-mounted WC cistern can be reached with a stepladder.

The storage tank in the loft is simply a container for cold water. Other than a leak, which is unlikely with modern tanks, the only problems that arise are caused by float-valve failure. The valve in a storage tank is similar to those used for WC cisterns, but you should never replace one with a miniature float valve.

Miniature float valve
This type of float valve is designed for installing in WC cisterns only.

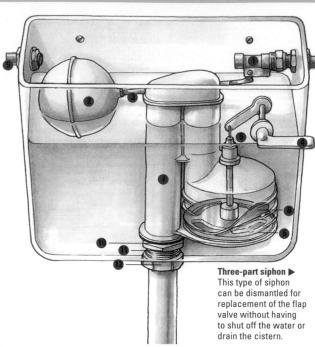

Direct-action cistern
The components of a typical direct-action WC cistern.
1 Float valve
2 Float
3 Float arm
4 Flushing lever
5 Wire link
6 Perforated plate
7 One-piece siphon
8 Flap valve
9 Overflow
10 Sealing washer
11 Retaining nut
12 Flush-pipe connector

Three-part siphon ▶
This type of siphon can be dismantled for replacement of the flap valve without having to shut off the water or drain the cistern.

1 Tie up the float arm

REPLACING A FLAP VALVE

If a WC cistern will not flush first time, take off the lid and check that the lever is actually operating the flushing mechanism. If that appears to be working normally, then try replacing the flap valve in the siphon. Before you service a one-piece siphon, shut off the water by tying the float arm to a batten placed across the cistern (**1**). Flush the cistern.

Use a large wrench to unscrew the nut that holds the flush pipe to the underside of the cistern (**2**). Move the pipe to one side.

Release the retaining nut that clamps the siphon to the base of the cistern (**3**). A little water will run out as you loosen the nut – so have a bucket handy. (You may find that the siphon is bolted to the base of the cistern, instead of being clamped by a single retaining nut.)

Disconnect the flushing arm, then ease the siphon out of the cistern. Lift the diaphragm off the metal plate (**4**) and replace it with one of the same size. Reassemble the entire flushing mechanism in the reverse order and reconnect the flush pipe to the cistern.

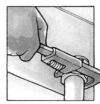

2 Release flush pipe

3 Loosen retaining nut

4 Lift off flap valve

Direct-action WC cisterns

Most modern WCs are washed down by means of direct-action cisterns. Water enters the cistern through a valve, which is opened and closed by the action of a hollow float attached to one end of a rigid arm. As the water rises in the cistern, it lifts the float until the other end of the arm closes the valve and shuts off the supply.

Flushing is carried out by depressing a lever, which is linked by wire to a rod attached to a perforated plastic or metal plate at the bottom of an inverted U-bend tube (siphon). As the plate rises, the perforations are sealed by a flexible plastic diaphragm (flap valve), so the plate can displace a body of water over the U-bend to promote a siphoning action. The water pressure behind the diaphragm lifts it, so that the contents of the cistern flow up through the perforations in the plate, over the U-bend and down the flush pipe. As the water level in the cistern drops, so does the float – thus opening the float valve to refill the cistern.

Servicing cisterns
The few problems associated with this type of cistern are easy to solve. A faulty float valve or poorly adjusted float arm will allow water to leak into the cistern until it drips from the overflow pipe that runs to the outside of the house. Slow or noisy filling can often be rectified by replacing the float valve. If the cistern will not flush until the lever is operated several times, the flap valve probably needs replacing (see left). If the flushing lever feels slack, check that the wire link at the end of the flushing arm is intact. When water runs continuously into the pan, check the condition of the washer at the base of the siphon.

Making a new wire link

It is impossible to flush a WC cistern if the flushing lever has come adrift.

You may find the old link is lying at the bottom of the cistern – but if not, you can bend one from a piece of thick wire. If you have thin wire only, twist the ends together with pliers to make a temporary repair.

Curing continuous running water

If you notice that water is running into the pan continuously, turn off the supply and let the cistern drain. Then check to see whether the siphon has split. If not, try changing the sealing washer.

Alternatively, the water may be flowing from the float valve so quickly that the siphoning action is not interrupted. The solution is to fit a float-valve seat with a smaller water inlet (see opposite).

☞ **SEE ALSO:** Adjusting the float arm 14, Spanners and wrenches 77–8

Renovating float valves

The pivoting end of the float arm on a diaphragm valve (known in the trade as a Part 2 valve) presses against the end of a small plastic piston, which moves the large rubber diaphragm to seal the water inlet.

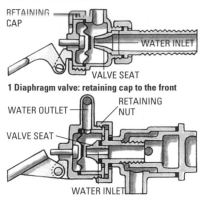

1 Diaphragm valve: retaining cap to the front

2 Diaphragm valve: retaining nut to the rear

Replacing the diaphragm

Turn off the water supply, then unscrew the large retaining cap. Depending on the model, the nut may be screwed onto the end of the valve (**1**) or behind it (**2**).

With the latter type of valve, slide out the cartridge inside the body (**3**) to find the diaphragm behind it. With the former, you will find a similar piston and diaphragm immediately behind the retaining cap (**4**).

Wash the valve, before assembling it along with the new diaphragm.

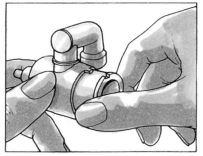

3 Slide out the cartridge to release the diaphragm

4 Undo the cap and pull float arm to find the valve

A faulty float valve is responsible for most of the difficulties that arise with WC cisterns and water-storage tanks. The water inlet inside the valve used to be sealed with a washer, whereas modern valves are fitted with a large diaphragm instead, designed to protect the mechanism from scale deposits. You can still obtain the earlier valves, but fit a diaphragm valve in a new installation.

If the inlet isn't sealed properly, water continues to feed into the cistern and escapes via the overflow. Some overflow pipes aren't able to cope with a full flow of mains water, so repair a dripping float valve before the flow becomes a torrent.

Servicing Portsmouth-pattern valves

In a Portsmouth-pattern valve, a piston moves horizontally inside the hollow metal body. The float arm, pivoting on a split pin, moves the piston back and forth to control the flow of water. A washer trapped in the end of the piston finally seals the inlet by pressing against the valve seat. If you have to force the valve closed to stop water dripping, it's time to replace the washer.

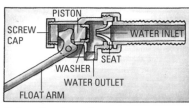

Portsmouth-pattern valve

Replacing the washer
Cut off the supply of water to the cistern or tank and flush the water out, in case you drop a component. Remove the split pin from beneath the valve and detach the float arm.

If there is a screw cap on the end of the valve body, remove it (**1**), using a pair of slip-joint pliers (you may have to apply a little penetrating oil to ease the threads). Insert the tip of a screwdriver in the slot beneath the valve body and slide the piston out (**2**).

To remove the washer, unscrew the end cap of the piston with pliers. Steady the piston by holding a screwdriver in its slot (**3**). Pick the old washer out of the cap (**4**) – but before replacing the washer,

clean the piston with fine wire wool.

Some pistons don't have a removable end cap, and so the washer has to be dug out with a pointed knife. Since it's a tight fit within a groove in the piston, make sure you don't damage this type of washer when replacing it.

Use wet-and-dry paper wrapped around a dowel rod to clean inside the valve body, but take care not to damage the valve seat at the far end.

Reassemble the piston and smear it lightly with silicone grease. Assemble the valve, then connect the float arm. Restore the supply of water and adjust the arm to regulate the water level in the cistern.

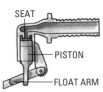

Croydon-pattern valve
Only old-fashioned tanks will be fitted with this valve. The piston travels vertically to close against the seat. Replace the washer as described left.

Interchangeable valve seats
The plastic seat against which the washer or diaphragm closes has a large inlet for low-pressure water or a small inlet for mains or high pressure. Seats that are damaged or worn should be replaced.

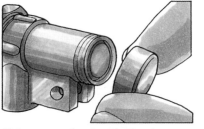

1 Take screw cap from the end of the valve

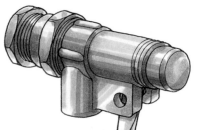

2 Slide the piston out with a screwdriver

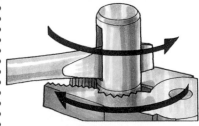

3 Split the piston into two parts

4 Pick out the washer with a screwdriver

☞ **SEE ALSO: Turning off the water 6–9, Adjusting a float arm 14, Slip-joint pliers 79**

Renovating valves and floats

Adjusting the float arm

Adjust the float so as to maintain the optimum level of water, which is about 25mm (1in) below the outlet of the overflow pipe.

The arm on a Portsmouth-pattern valve is usually a solid-metal rod. You bend it downward slightly to reduce the water level, or straighten it in order to admit more water (**1**).

The arm on a diaphragm valve has an adjusting screw, which presses on the end of the piston. Release the lock nut and turn the screw towards the valve to lower the water level, or away from it to allow the water to rise (**2**).

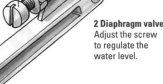

1 Straighten or bend a metal float arm

Thumb-screw adjustment
Some float arms are cranked, and the float is attached with a thumb-screw clamp. To adjust the water level in the cistern, slide the float up or down the rod.

Float valve with flexible silencer tube

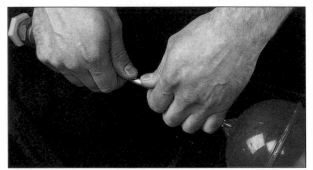

SCREW REGULATOR

2 Diaphragm valve
Adjust the screw to regulate the water level.

Replacing the float

Modern plastic floats rarely leak, but old-style metal floats eventually corrode and allow water to seep into the ball. The float gradually sinks, until it won't ride high enough to close the valve.

Unscrew the float and shake it, to find out whether there is water inside.

If you won't be able to obtain a new float for several days, lay the ball on a bench, enlarge the leaking hole with a screwdriver and pour out the water. Cover the ball with a plastic bag, tying the neck tightly around the float arm, and then replace the float.

Curing noisy cisterns

Cisterns that fill noisily can be very annoying, particularly if the WC is situated right next to a bedroom. It was once permitted to screw a pipe into the outlet of a valve so that it hung vertically below the level of the water. This solved the problem of water splashing into the cistern, but water companies were concerned about the possibility of water 'back-siphoning' through the silencer tube into the mains supply. Although rigid tubes are banned nowadays, you are permitted to fit a valve with a flexible plastic silencer tube (see far left), because it will seal itself by collapsing should back-siphoning occur.

A silencer tube can also prevent water hammer – a rhythmic thudding that reverberates along the pipework. This is often the result of ripples on the surface of the water in a cistern, caused by a heavy flow from the float valve. As the water rises, the float arm bouncing on the ripples 'hammers' the valve, and the sound is amplified and transmitted along the pipes. A flexible plastic tube will eliminate ripples by introducing water below the surface.

If the water pressure through the valve is too high, the arm oscillates as it tries to close the valve – another cause of water hammer. This can be cured by fitting an equilibrium valve. As water flows through the valve, some of it is introduced behind the piston or diaphragm to equalize the pressure on each side, so that the valve closes smoothly and silently.

Before swapping your present valve, check that the pipework is clipped securely – as the noise could be caused by vibrating pipes.

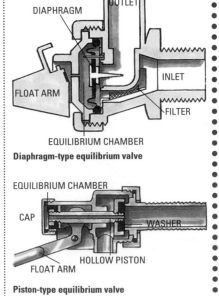

Diaphragm-type equilibrium valve

Piston-type equilibrium valve

Renewing a float valve

Turn off the supply of water to the cistern or tank and flush the pipework, then use a spanner to loosen the tap connector joining the supply pipe to the float-valve stem. Remove the float arm, then unscrew the fixing nut on the outside of the cistern and pull out the valve.

Fit the replacement valve and, if possible, use the same tap connector to join it to the supply pipe. Adjust and tighten the fixing nuts to clamp the new valve to the cistern, then turn the water supply back on and adjust the float arm.

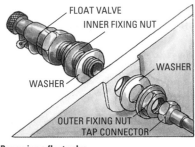

Renewing a float valve
Clamp the valve to the cistern with fixing nuts.

CHOOSING THE CORRECT PRESSURE

Float valves are made to suit different water pressures: low, medium and high (LP, MP and HP). It is important to choose a valve of the correct pressure, or the cistern may take a long time to fill. Conversely, if the water pressure is too high for the valve, it may leak continuously. Those fed direct from the mains should be HP valves, whereas most domestic WC cisterns require an LP valve. If the head (the height of the tank above the float valve) is greater than 13.5m (45ft), fit an MP valve. In those rare cases where the head exceeds 30m (100ft), fit an HP valve. In an apartment with a packaged plumbing system (a storage tank built on top of the hot-water cylinder), the pressure may be so low that you will have to fit a full-way valve to the WC cistern in order to get it to fill reasonably quickly. If you live in an area where water pressure fluctuates a great deal, fit an equilibrium valve (see left).

To alter the pressure of a modern valve, simply replace the seat inside it. If the valve is a very old pattern, you will have to swap it for another one of a different pressure.

☛ SEE ALSO: Float valves 8, 13, 81, Supporting pipes 23

Drainage systems

A drainage system is designed to carry dirty water and WC waste from the appliances in your home to underground drains leading to the main sewer. The various branches of the waste system are protected by U-bend traps full of water, to stop drain smells fouling the house. Depending on the age of your house, it will have a two-pipe system or a single stack. Because the two-pipe system has been in use for very much longer, it is still the more common of the two. Use similar methods to maintain either system.

Two-pipe system

The waste pipes of older houses are divided into two separate systems. WC waste is fed into a large-diameter vertical soil pipe that leads directly to the underground drains. To discharge drain gases at a safe height and make sure that back-siphoning cannot empty the WC traps, the soil pipe is vented to the open air above the guttering.

Individual branch pipes leading from upstairs washbasins and baths drain into an open hopper that funnels the water into another vertical waste pipe. Instead of feeding directly into the underground drains, this pipe terminates over a yard gully – another trap covered by a grid. A separate waste pipe from the kitchen sink normally drains into the same gully.

The yard gully and soil pipe both discharge into an underground inspection chamber, or manhole. These chambers provide access to the main drains for clearing blockages, and there will be one wherever your main drain changes direction on its way to the sewer.

At the last inspection chamber, just before the drain enters the sewer, there is an interceptor trap, the final barrier to drain gases and sewer rats.

Single-stack system

Since the late 1950s, most houses have been drained using a single-stack system. Waste from basins, baths and WCs is fed into the same vertical soil pipe or stack – which, unlike the two-pipe system, is often built inside the house. A single-stack system must be designed carefully to prevent a heavy discharge of waste from one appliance siphoning the trap of another, and to avoid the possibility of WC waste blocking other branch pipes. The vent pipe of the stack terminates above the roof and is capped with an open cage; or inside the house and is fitted with an air-admittance valve (see far right).

The kitchen sink can be drained through the same stack, but it is still common practice to drain sink waste into a yard gully. Nowadays waste pipes must pass through the grid, stopping short of the water in the gully trap – so that even if blocked with leaves, the waste can discharge unobstructed into the gully. Alternatively, it can be a back-inlet gully, with the waste pipe entering below ground level.

A downstairs WC is sometimes drained through its own branch drain to an inspection chamber.

RESPONSIBILITY FOR DRAINS

If a house is drained individually, the whole system up to the point where it joins the sewer is the responsibility of the householder. However, where a house is connected to a communal drainage system linking several houses, the arrangement for maintenance, including the clearance of blockages, is not so straightforward.

If the drains were constructed prior to 1937, the local council is responsible for cleansing but can reclaim the cost of repairing any part of the communal system from the householders. After that date, all responsibility falls upon the householders collectively, so that they are required to share the cost of the repair and cleansing of the drains up to the sewer, no matter where the problem occurs. Contact the Technical Services Department of your local council to find out who is responsible for your drains.

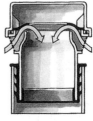

Ventilating pipes and stacks
An air-admittance valve seals off the vent pipe, but allows air into the system to prevent water being siphoned from the trap seals. This type of valve can only be used if the drainage scheme has been approved by the local authority.

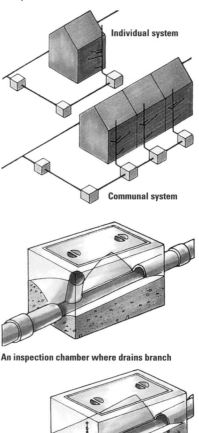

Individual system

Communal system

An inspection chamber where drains branch

A chamber with interceptor trap

Prefabricated chamber
On a modern drainage system, the inspection chambers may take the form of cylindrical prefabricated units. There may not be an interceptor trap in the last chamber before the sewer.

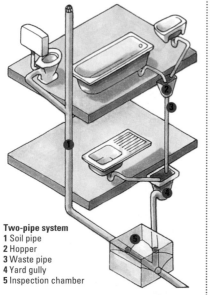

Two-pipe system
1 Soil pipe
2 Hopper
3 Waste pipe
4 Yard gully
5 Inspection chamber

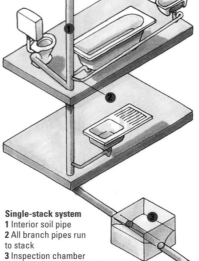

Single-stack system
1 Interior soil pipe
2 All branch pipes run to stack
3 Inspection chamber

☞ SEE ALSO: Plumbing systems 6–8, Blocked soil pipe 17, Yard gully 17, Blocked drains 18

Clearing blocked sinks and basins

Don't ignore the early signs of an imminent blockage in the waste pipe from a sink, bath or basin. If the water drains away slowly, use a chemical cleaner to remove a partial blockage before you are faced with clearing a serious obstruction. If a waste pipe blocks without warning, try a series of measures to locate and clear the obstruction.

Cleansing the waste pipe

Grease, hair and particles of kitchen debris build up gradually within the traps and waste pipes. Regular cleaning with a proprietary chemical drain cleaner will keep the waste system clear and sweet-smelling.

If water drains away sluggishly, use a cleaner immediately. Follow the manufacturer's instructions carefully, with particular regard to safety. Always wear protective gloves and goggles when handling chemical cleaners, and keep them out of the reach of children.

If unpleasant odours linger after you've cleaned the waste, pour a little disinfectant into the basin overflow.

Using a plunger

If one basin fails to empty while others are functioning normally, the blockage must be somewhere along its individual branch pipe. Before you attempt to locate the blockage, try forcing it out of the pipe with a sink plunger. Smear the rim of the rubber cup with petroleum jelly, then lower it into the blocked basin to cover the waste outlet. Make sure that there's enough water in the basin to cover the cup. Hold a wet cloth in the overflow with one hand while you pump the handle of the plunger up and down a few times. The waste may not clear immediately if the blockage is merely forced further along the pipe, so repeat the process until the water drains away. If it will not clear after several attempts, try clearing the trap, or use a pump to clear the pipe (see left).

Clearing the trap

The trap situated immediately below the waste outlet of a sink or basin is basically a bent tube designed to hold water to seal out drain odours. Traps become blocked when debris collects at the lowest point of the bend.

Place a bucket under the basin to catch the water, then use a wrench to release the cleaning eye at the base of a standard trap; on a bottle trap, remove the large access cap by hand. If there is no provision for gaining access to the trap, unscrew the connecting nuts and remove the entire trap.

Let the contents of the trap drain into the bucket, then bend a hook on the end of a length of wire and use it to probe the section of waste pipe beyond the trap. (It is also worth checking outside, to see if the other end of the pipe is blocked with leaves.) If you have had to remove the trap, take the opportunity to scrub it out with detergent before replacing it.

Cleaning the branch pipe

Quite often, a vertical pipe from the trap joins a virtually horizontal section of the waste pipe. There should be an access plug built into the joint, so that you can clear the horizontal pipe. Have a bowl ready to collect any trapped water, then unscrew the plug by hand. Use a length of hooked wire to probe the branch pipe. If you locate a blockage that seems very firmly lodged, rent a drain auger from a tool-hire company to clear the pipework.

If there's no access plug, remove the trap and probe the waste pipe with an auger. If the pipe is constructed with push-fit joints, you can dismantle it.

Using a pump
Block the sink overflow with a wet cloth. Fill the pump with water from the tap, then hold its nozzle over the outlet, pressing down firmly. Pump up and down until the obstruction is cleared.

USING A PUMP TO CLEAR A BLOCKAGE

If a plunger is ineffective in clearing a blocked waste outlet, use a simple hand-operated hydraulic pump. A downward stroke on the tool forces a powerful jet of water along the pipe to disperse the blockage. If the blockage is lodged firmly, an upward stroke creates enough suction to pull it free.

Use a plunger to force out a blockage

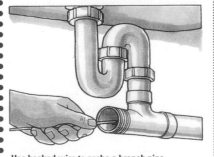

Use hooked wire to probe a branch pipe

Unscrew the access cap on a bottle trap

Tubular trap
If the access cap to the cleaning eye is stiff, use a wrench to remove it.

Bottle trap
This type of trap can be cleared easily because the whole base of the trap unscrews by hand.

☛ **SEE ALSO:** Frozen pipes 9, Plungers 16, 74, Drain auger 17, 74

If several fittings are draining poorly, the vertical stack is probably obstructed. In autumn, the hopper, downpipe and yard gully may be blocked with leaves. The blockage may not be obvious when you empty a basin, but the contents of a bath will almost certainly cause an overflow. Clear the blockage urgently to avoid penetrating damp.

Cleaning out the hopper and drainpipe

Wearing protective gloves, scoop out the debris from the hopper, then gently probe the drainpipe with a cane to check that it is free. Clear the bottom end of the pipe with a piece of bent wire. If an old cast-iron waste pipe has been replaced with a modern plastic pipe, you may find there are cleaning eyes or access plugs at strategic points for clearing a blockage.

While you're on the ladder, scrub the inside of the hopper and disinfect it to prevent stale odours entering a nearby bathroom.

Unblocking a yard gully

Unless you decide to hire an auger, you have little option but to clear a blocked gully by hand. However, by the time it overflows the water in the gully will be quite deep, so try bailing some of it out with a small disposable container. Wearing rubber gloves, scoop out the debris from the trap until the remaining water disperses.

Rinse the gully with a hose and cleanse it with disinfectant. Scrub the grid as clean as possible, or burn off accumulated grime from a metal grid with a gas torch.

If a flooded gully appears to be clear and yet the water will not drain away, try to locate the blockage at the nearest inspection chamber.

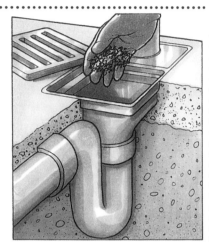

Bail out the water, then clear a gully by hand

Unblocking a soil pipe

Unblocking a soil pipe is an unpleasant job and it's worth hiring a professional cleaning company – especially if the pipe is made of cast iron, as it will almost certainly have to be cleared via the vent above the roof.

You can clean a modern plastic stack yourself, since there should be a large hinged cleaning eye, or other access plugs, wherever branch pipes join the stack. If the stack is inside the house, lay polythene sheets on the floor and be prepared to mop up trapped sewage when it spills from the pipe.

Unscrew and open the cleaning eye to insert a hired drain auger. Pass the auger into the stack until you locate the obstruction, then crank the handle to engage it. Push or pull the auger until you can dislodge the obstruction to clear the trapped water, then hose out the stack. Wash and disinfect the surrounding area.

If the water in a WC pan rises when you flush it, there's a blockage in the vicinity of the trap. A partial blockage allows the water level to fall slowly.

Hire a larger version of the sink plunger to force the obstruction into the soil pipe. Position the rubber cup of the plunger well down into the U-bend, and pump the handle. When the blockage clears, the water level will drop suddenly, accompanied by an audible gurgling.

If the trap is blocked solidly, hire a special WC auger. Pass the flexible clearing rod as far as possible into the trap, then crank the handle to dislodge the blockage. Wash the auger in hot water and disinfect it, before returning it to the hire company.

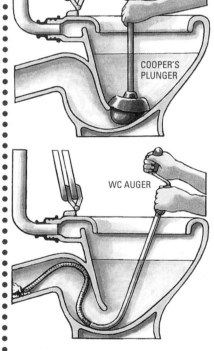

COOPER'S PLUNGER

WC AUGER

Clearing a blockage
Use a Cooper's plunger (left) to pump a blocked WC. Alternatively, clear it with a special WC auger (below left).

● **Clearing a blockage with a hydraulic pump**
Shift a really stubborn blockage with a hired pump, similar to the one used for clearing a blocked sink (see opposite)

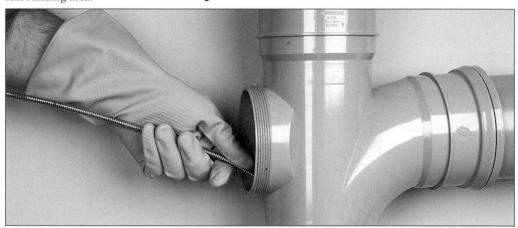

Use a hired auger to clear a soil stack

Rodding the drains

The first sign of a blocked drain could be an unpleasant smell from an inspection chamber, but a severe blockage may cause sewage to overflow from a gully or from beneath the cover of an inspection chamber. Before you resort to professional services, hire a set of drain rods – short flexible rods made of plastic or wire, screwed end to end – to clear the blockage.

Locating the blockage

Lift the cover from the inspection chamber nearest to the house. If it's stuck or the handles have rusted away, scrape the dirt from around its edges and prise it up with a garden spade.

● If the chamber contains water, check the one nearer the road or boundary. If that chamber is dry, the blockage is between the two chambers.

● If the chamber nearest the road is full, the blockage will be in the interceptor trap or in the pipe beyond, leading to the sewer.

● If both chambers are dry and yet either a yard gully or downstairs WC will not empty, check for blockages in the branch drains that run to the first inspection chamber.

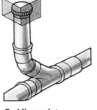

Rodding points
A modern drainage system is often fitted with rodding points to provide access to the drain. They are sealed with small oval or circular covers.

Rodding the drainpipe

Screw two or three rods together and attach a corkscrew fitting to the end. Insert the rods into the drain at the bottom of the inspection chamber, in the direction of the suspected blockage. If the chamber is full of water, use the end of a rod to locate the open channel running across the floor, leading to the mouth of the drain.

As you pass the rods along the pipe, attach further lengths till you reach the obstruction, then twist the rods clockwise to engage the screw. (Never twist the rods anticlockwise, or they will become detached.) Pull and push the obstruction until it breaks up, allowing the water to flow away.

Extract the rods, flush the chamber with clean water from a hose, and then replace the lid.

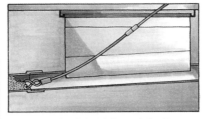

Use a corkscrew fitting to clear a drain

Cesspools and septic tanks

Houses built in the country or on the outskirts of a town are not always connected to a public sewer. Instead, waste is drained into a cesspool or septic tank.

A cesspool simply acts as a collection point for sewage until it can be pumped out by the local council – whereas a septic tank is a complete waste-disposal system, in which sewage is broken down by bacterial action before the water is finally discharged into a local waterway or distributed underground.

Cesspools

The Building Regulations stipulate that cesspools must have a minimum capacity of 18cu m (4000 gallons), but many existing cesspools accommodate far less and require emptying perhaps once every two weeks. Before buying a country home with a cesspool, it is worth checking that it will cope with your needs. Water authorities estimate the disposal of approximately 115 litres (25 gallons) per person per day.

Most cesspools are cylindrical pits lined with brick or concrete. Modern ones are sometimes prefabricated in glass-reinforced plastic. Access is via a manhole cover.

Clearing interceptor traps

Screw a rubber plunger to the end of a short length of rods and locate the channel that leads to the base of the trap. Push the plunger into the opening of the trap, then pump the rods a few times to expel the blockage. (This is also a useful technique for clearing blocked yard gullies.)

If the water level does not drop after several attempts, try clearing the drain leading to the sewer. Access to this drain is through a cleaning eye above the trap. It will be sealed with a stopper, which you will have to dislodge with a drain rod, unless it is attached to a chain stapled to the chamber wall. Don't let the stopper fall into the channel and block the trap. Rod the drain to the sewer, then hose out the chamber before replacing the stopper and cover.

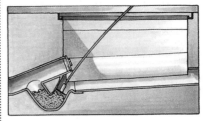

To rod an interceptor trap, fit a rubber plunger

Septic tanks

The sewage in a septic tank separates slowly: heavy sludge falls to the bottom to leave relatively clear water, with a layer of scum floating on the surface. A dip-pipe discharges waste below the surface, so that incoming water does not stir up the sewage. Bacterial action takes a minimum of 24 hours, so the tank is divided into chambers by baffles to slow down the movement of sewage through the tank.

The partly treated waste passes out of the tank, through another dip-pipe, into some form of filtration system that allows further bacterial action to take place. This may consist of another chamber, containing a deep filter bed; or the waste may flow underground through a network of drains, which disperses the water over a wide area to filter through the soil.

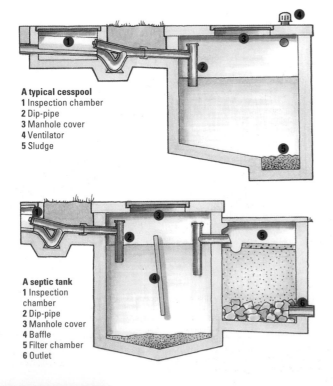

A typical cesspool
1 Inspection chamber
2 Dip-pipe
3 Manhole cover
4 Ventilator
5 Sludge

A septic tank
1 Inspection chamber
2 Dip-pipe
3 Manhole cover
4 Baffle
5 Filter chamber
6 Outlet

☛ **SEE ALSO:** Inspection chambers 15, Drain rods 74

Metal pipes

The ability to install a run of pipework, make watertight joints and connect up to fittings constitutes the basis of most plumbing. Without these skills, a householder is restricted to simple maintenance. Modern materials and technology have made it possible for anybody who is prepared to master a few techniques to upgrade and extend plumbing without having to hire a professional.

Metric and imperial pipes

Copper and stainless-steel pipes are now made in metric sizes, whereas pipework already installed in older house will have been made to imperial measurements. If you compare the equivalent dimensions (15mm – ½in, 22mm – ¾in, 28mm – 1in), the difference seems obvious, but metric pipe is measured externally while imperial pipe is measured internally. In fact, the difference is very small – but enough to cause some problems when joining one type of pipe to the other.

When making soldered joints, an exact fit is essential. Imperial to metric adaptors are necessary when joining 22mm pipe to its imperial equivalent; and, although not essential, adaptors are convenient when you are working with 28mm pipes or with thick-walled ½in pipes. Adaptors are not required when using compression fittings, but when you are connecting 22mm to ¾in plumbing slip an imperial olive onto the ¾in pipe.

Typically, 15mm (½in) pipe is used for the supply to basins, kitchen sinks washing machines, some showers, and radiator flow and returns. However, 22mm (¾in) pipes are used to supply baths, high-output showers, hot-water cylinders and main central-heating circuits; and 28mm (1in) pipe for larger heating installations.

Electrochemical action

Joining pipes made from different metals can accelerate corrosion as a result of electrolytic action. If you live in a soft-water area, where this problem tends to be pronounced, use plastic pipe and connectors when you're joining to old pipework – but make sure that the metal pipes are still bonded to earth, as required by the Wiring Regulations.

Metal supply pipes

Over the years, most household plumbing systems will have undergone some form of improvement or alteration. As a result, you may find any of a number of metals used, perhaps in combination, depending on the availability of materials at the time of installation or the preference of an individual plumber.

Copper

Half-hard-tempered copper tubing is by far the most widely used material for pipework. This is because it's lightweight, solders well, and can be bent easily (even by hand, with the aid of a bending spring). It is employed for both hot-water and cold-water pipes, as well as for central-heating systems. There are three sizes of pipe that are invariably used for general domestic plumbing: 15mm (½in), 22mm (¾in), and 28mm (1in).

Stainless steel

Stainless-steel tubing is not as common as copper, but is available in the same sizes. You may have to order it from a plumbers' merchant. It's harder than copper, so cannot be bent as easily, and is difficult to solder. It pays to use compression joints to connect stainless-steel pipes, but tighten them slightly more than you would when joining copper.

Stainless steel does not react with galvanized steel (iron) – see ELECTRO-CHEMICAL ACTION (bottom left).

Lead

Lead is never used for any form of new plumbing – but there are thousands of houses that still have a lead rising main connected to a modernized system.

Lead plumbing that's still in use must be nearing the end of its life, so replace it as soon as an opportunity arises. When drinking water lies in a lead pipe for some time, it absorbs toxins from the metal. If you have a lead pipe supplying your drinking water, always run off a little water before you use any.

Galvanized steel (iron)

Galvanized steel was once commonly used for supply pipes, both below and above ground, having taken over from lead. It was then superseded by copper.

There are two problems with this type of pipe. It rusts from the inside and resists water flow as it deteriorates. Also, when it is joined to copper, the galvanizing breaks down rapidly because of an electrolytic action between the copper and zinc coating (see bottom left).

● **Cast-iron waste pipes**
All old soil pipes are made from cast iron, which is prone to rusting. If it weren't for their relatively thick walls, pipes of this kind would have rusted away long ago.

Plastic waste pipes
Should you need to replace a cast-iron pipe, ask for one of the plastic alternatives.

Copper pipes
The economic choice for modern, plumbing systems.

Stainless steel
Due to its superior appearance and strength, stainless steel is sometimes used where pipe runs are exposed. It does not cause electrolytic action with galvanised-steel pipes.

Lead
This is still found in older houses. It can introduce toxins into the drinking-water supply, so should be replaced.

Iron
Iron pipes are used for mains water supply in some older systems. Iron is susceptible to furring-up and decay, which can result in low water pressure and leaks. Cast iron is used for waste pipes in older buildings.

☛ SEE ALSO: Soldered joints 21, Bending springs 23, Push-fit joints 25–6, Plastic waste pipes 26, Soft water 48, Main switch equipment 68, Supplementary bonding 69–70

Metal joints and fittings

Joints are made to connect pipes at different angles and in various combinations. There are adaptors for joining metric and imperial pipes, and for connecting one kind of material to another. You need to consult manufacturers' catalogues to see every variation, but the examples on this page illustrate a typical range of joints.

Plumbing fittings such as valves are made with demountable compression joints, so that they can be removed easily for servicing or replacement.

Pipe joints

Corrosion resistance
Corrosion can take place between brass fittings and copper pipes. Look for the symbol that denotes corrosion-resistant brass fittings.

Soldering capillary joints
Solder is introduced to each mouth of the assembled end-feed joint (far right) and flows by capillary action into the fitting.

The rings pressed into the sleeves of an integral-ring fitting (right) contain the exact amount of solder to make perfect joints.

It would be impossible to make strong, watertight joints by simply soldering two lengths of copper pipe end to end. Instead, plumbers use capillary or compression joints.

Capillary joints
Capillary joints are made to fit snugly over the ends of a pipe. The very small space between the pipe and joint sleeve is filled with molten solder. When it solidifies on cooling, the solder holds the joint together and makes it watertight. Capillary joints are neat and inexpensive – but because you need to heat the metal with a gas torch, there is a slight risk of fire when working in confined spaces under floors.

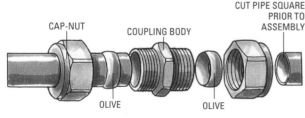

CAP-NUT
COUPLING BODY
CUT PIPE SQUARE PRIOR TO ASSEMBLY
OLIVE
OLIVE

Compression joints
Compression joints are very easy to use, but are more expensive than capillary joints. They are also more obtrusive, and you will find it impossible to manoeuvre a wrench where space is restricted. The end of each pipe is cut square before the joint is assembled. When the cap-nut is tightened with a wrench it compresses a ring of soft metal, known as an olive, to fill the joint between fitting and pipe.

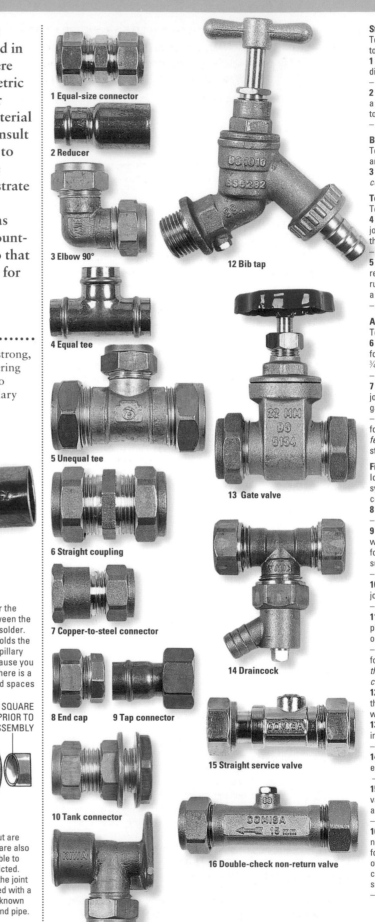

1 Equal-size connector

2 Reducer

3 Elbow 90°

4 Equal tee

5 Unequal tee

6 Straight coupling

7 Copper-to-steel connector

8 End cap

9 Tap connector

10 Tank connector

11 Bib-tap wall plate

12 Bib tap

13 Gate valve

14 Draincock

15 Straight service valve

16 Double-check non-return valve

Straight connectors
To join two pipes end to end in a straight line.
1 For pipes of equal diameter
– compression joint.
2 Reducer to connect a 22mm (¾in) pipe to a 15mm (½in) pipe
– capillary joint.

Bends or elbows
To join two pipes at an angle.
3 Elbow 90° –
compression joint.

Tees (T-joints)
To join three pipes.
4 Equal tee, for joining three pipes of the same diameter
– capillary joint.
5 Unequal tee, for reducing size of pipe run when connecting a branch pipe
– compression joint.

Adaptors
To join dissimilar pipes.
6 Straight coupling for joining 22mm and ¾in pipes
– compression joint.
7 Connector for joining copper to galvanized steel
– compression joint for copper, threaded female coupling for steel.

Fittings
Identical jointing systems are used to connect fittings.
8 End cap, to seal pipes
– compression joint.
9 Tap connector, with threaded nut for connecting supply pipe to tap
– capillary joint.
10 Tank connector, joins pipes to cisterns
– compression joint.
11 Bib-tap wall plate, for fixing tap on outside wall
– compression joint for supply pipe, threaded female connector for tap.
12 Bib tap has threaded tail to fit wall plate.
13 Gate valve to fit in straight pipe run
– compression joint.
14 Draincock for emptying a pipe run
– compression joint.
15 Straight service valve for isolating a tap or float valve
– compression joint.
16 Double-check non-return valve, used for outside taps and other outlets where contamination of water supply is possible
– compression joint.

Making soldered joints

Calculate the length of pipe you need, allowing enough to fit into the sleeve of the joint at each end. Whatever type of joint you use, it's essential to cut the end of every length of pipe square.

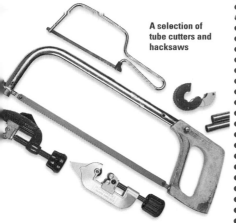

A selection of tube cutters and hacksaws

To ensure a perfectly square cut each time, use a tube cutter. Align the cutting wheel with your mark, and adjust the handle of the tool to clamp the rollers against the pipe **(1)**. Rotate the tool around the pipe, adjusting the handle after each revolution to make the cutter bite deeper into the metal.

A tube cutter makes a clean cut on the outside of the pipe, but use the pointed reamer on the tool to clean the burr from inside the cut end **(2)**.

If you use a hacksaw, make sure the cut is square by wrapping a piece of paper with a straight edge around the pipe. Align the wrapped edge and use it to guide the saw blade **(3)**. Remove the burr, inside and out, with a file.

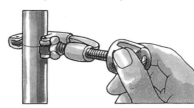

1 Clamp the tube cutter onto the pipe

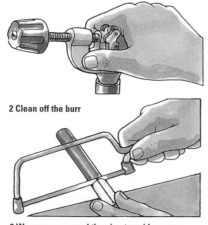

2 Clean off the burr

3 Wrap paper around the pipe to guide a saw

Soldering pipe joints is easy once you have had a little practice. The fittings are cheap, so you can afford to try out the techniques before you begin to install pipework. You need a gas torch to apply heat, some flux to clean the metal, and solder to make the joint. Make sure the pipe is perfectly dry before you attempt to solder a joint.

Gas torches

To heat the metal sufficiently for a soldered joint, most plumbers use a gas torch. Gas, liquefied under pressure, is contained in a disposable metal canister. When the control valve of the torch is opened, gas is vaporized to combine with air, making a highly combustible mixture. Once ignited, the flame is adjusted until it burns steadily with a clear blue colour.

Many professional plumbers use a propane torch, which is connected by a hose to a metal gas bottle. The average householder doesn't need such expensive equipment, but if you happen to own a propane torch, perhaps for car repairs, you can use the same tool for soldering plumbing joints.

Using integral-ring joints

Clean the ends of each pipe and the inside of the joint sleeves with wire wool or abrasive paper until the metal is shiny. Brush flux onto the cleaned metal and push the pipes into the joint, twisting them to spread the flux evenly. Push each pipe up against the stop in the joint.

If you are using elbows or tees, mark the pipe and joint with a pencil, to make sure they do not get misaligned during the soldering.

Slip a ceramic tile or a plumber's fibreglass mat behind the joint to protect flammable materials, then apply the flame of a gas torch to the area of the joint to heat it evenly. When a bright ring of solder appears at each end of the joint, remove the flame and allow the metal to cool for a couple of minutes before disturbing it.

Repairing a weeping joint
When you fill a new installation with water for the first time, check every joint to make sure it's watertight. If you notice water 'weeping' from a soldered joint, drain the pipe and allow it to dry. Heat the joint and apply some fresh solder to the edge of each mouth. If it leaks a second time, heat the joint until you can pull it apart with gloved hands. Either use a new joint or clean and flux all surfaces and reuse the same joint, adding solder as if you were working with an end-feed fitting (see right).

Solder and flux

Solder is a soft alloy manufactured with a melting point lower than that of the metal it is joining. Plumbers' solder is sold as wound wire.

Copper must be spotlessly clean and grease-free if it is to produce a properly soldered joint. Even when you have cleaned it mechanically with wire wool, copper begins to oxidize immediately; a chemical cleaner known as flux is therefore painted onto the metal to provide a barrier against oxidation until the solder is applied. A non-corrosive flux in the form of a paste is the best one to use. On stainless-steel pipework use a highly efficient active flux – but wash it off with warm water after the joint is made, or the metal will corrode.

Using end-feed joints

Having cleaned and assembled an end-feed joint, heat the area of the joint evenly. When the flux begins to bubble, remove the flame and touch the solder wire to two or three points around the mouth of each sleeve – the joint is full of solder when a bright ring appears around each sleeve. Allow it to cool.

Heat the joint to melt the captive solder

Introduce solder to a heated end-feed joint

● **Joining stainless-steel pipes**
The techniques for joining copper and stainless-steel are similar – but because the steel is harder, you will find that it's easier to cut it with a hacksaw. Use an active flux when soldering stainless steel (see left).

Gas torches
A gas torch is used for heating soldered joints. A simple torch (above) is available from any DIY outlet.
The propane torch (below) is used by professional plumbers.

● **Lead-free solder**
Use lead-free solder when joining pipes that will supply drinking water.

☛ **SEE ALSO: Pipe fittings 19, 24, 26, Plumbing tools 74–9**

Compression joints

Using compression fittings is so straightforward that you will be able to make watertight joints without any previous experience.

Assembling a joint

Cut the ends of each pipe square and clean them, along with the olives, using wire wool. Dismantle a new joint and slip a cap-nut over the end of one pipe, followed by an olive (1). Look carefully to see if the sloping sides of the olive are equal in length. If one is longer than the other, that side should face away from the nut.

Push the pipe firmly into the joint body (2), twisting it slightly to ensure it is firmly against the integral stop. Slide the olive up against the joint body, then tighten the nut by hand.

The olive must be compressed by just the right amount to ensure a watertight joint. As a guide, make a pencil mark on one face of the nut and on the opposing face on the joint body (3); then, holding the joint body steady with a spanner, use another spanner to turn the nut one complete revolution (4). Assemble the other half of the joint in exactly the same manner.

Some plumbers like to wrap a single turn of PTFE tape over the olive before tightening the nut, to make absolutely sure the joint is watertight. However, a properly tightened compression joint should be watertight without it.

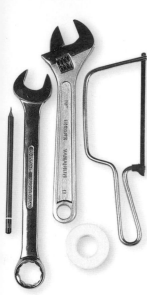

Straight connector
Compression joint to join two pipes of equal diameter, end to end, in a straight line.

1 Slip an olive onto the pipe after the cap-nut

2 Clamp the joint to the pipe with the nut

Elbow joint
A 90-degree elbow compression joint connects two pipes at an angle.

3 Mark the nut and joint with a pencil

4 Tighten the joint with two spanners

Repairing a weeping joint

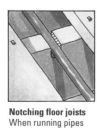

Notching floor joists
When running pipes under floorboards, notch each joist to receive the pipe. Cut the notch to align with the centre of a floor-board and drive a nail on each side when replacing the board.

Having filled the pipe with water, check each joint for leaks. Make one further quarter turn on any nut that appears to be weeping.

Crushing an olive by overtightening a compression joint will cause it to leak. Drain the pipe and dismantle the joint. Cut through the damaged olive with a junior hacksaw, taking care not to damage the pipe. Remake the joint with a new olive, restore the supply of water, and check for leaks once more.

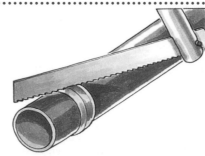

Saw through a damaged olive

Galvanized-steel pipe is connected by threaded joints, so if you plan to extend old pipework using the same material you will need a pipe die to cut the threads on the end of each length of new pipe.

You can hire a pipe die, but a simpler solution is to continue the run in plastic, using an adaptor to connect one system to another. One end of the adaptor has a push-fit sleeve for the plastic pipework; the other end has a male or female threaded connector for the galvanized steel.

Fitting an adaptor

Use two Stillson wrenches to unscrew the joint on the old pipework where you intend to connect up to plastic. Grip the joint with one wrench and the pipe with the other, pushing and pulling in the direction the jaws face (1). If the joint is stiff, use penetrating oil or play the flame of a gas torch along it.

Threaded connections leak unless they're made watertight with plumbers' PTFE tape. Wrap the tape clockwise two or three times around the pipe to cover the threads (2), then engage and tighten the adaptor.

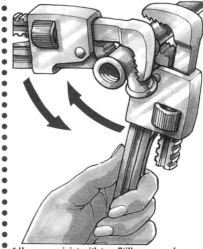

1 Unscrew a joint with two Stillson wrenches

2 Wrap plenty of PTFE tape over the threads

☞ SEE ALSO: Metal joints and fittings 20, Adaptors 20, Wrenches 78

When replacing old lead plumbing with copper, plumbers used to make the connection to the lead rising main with solder and a blowlamp. It is illegal to make such joints nowadays – and it is also much simpler to use a special lead-to-copper compression joint.

There are joints for connecting lead pipes to 15 and 22mm (½ and ¾in) copper pipes. You can use similar joints for plastic plumbing, provided you reinforce the plastic pipe with metal inserts. Although the connectors are specified according to the bore of lead pipework, measure the outside diameter of your rising main and ask a plumbers' merchant to provide a suitable compression joint.

Making the connection
Select a straight length of lead pipe that is as round as possible. It must also be in good condition, as the O-ring inside the fitting won't make a watertight seal if the lead is dented or scored.

Turn off the water supply. Have a bucket ready to catch the water, then cut the lead pipe with a hacksaw. Chamfer the outside edge of the pipe, and remove the burr from the inside. Dismantle the compression joint and check that the large thrust nut makes a good sliding fit on the lead pipe. You can scrape back a slightly oversize pipe to fit, keeping it as round as possible.

Slide the thrust nut onto the pipe, then the two metal rings and the rubber O-ring **(1)**. Slide the threaded coupling body onto the end of the pipe and push it against the internal end stop. Tighten the coupling **(2)** until you feel resistance, but don't use excessive force.

The other end of the coupling body carries a conventional compression joint for the copper pipe.

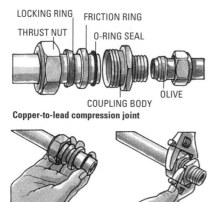

LOCKING RING   FRICTION RING
THRUST NUT   O-RING SEAL
                                OLIVE
COUPLING BODY
Copper-to-lead compression joint

1 Fit nut and rings   **2 Tighten the coupling**

You can change the direction of a pipe run by using an elbow joint, but there are occasions when bending the pipe itself will produce a neater or more accurate result.

If you want to carry a pipe over a small obstruction (another pipe, for example), a slight kink in the pipe will be less of an obstruction to the flow of water and will therefore create less noise than two elbows within a few centimetres of each other. It is also cheaper.

Perhaps you want to run pipes into a window alcove where the walls meet at an unusual angle? Bending the pipes accurately will allow you to fit the pipes neatly against the alcove walls.

Using a bending spring

A bending spring is the cheapest and easiest tool for making bends in small pipe runs. It is a hardened-steel coil spring that supports the walls of copper tube to stop it kinking. Most bending springs are made to fit inside the pipe, but some slide over it.

Slide the spring into the tube, so it supports the area you want to bend. Hold the tube against your padded knee and bend it to the required angle. The bent tube will grip the spring, but slipping a screwdriver into the ring at one end and turning it anticlockwise will reduce the diameter of the spring so that you can pull it out.

If you make a bend some distance from the end of a tube, you won't be able to withdraw the bending spring in the normal way. Either use an external spring or tie a length of twine to the ring and lightly grease the spring with petroleum jelly before you insert it. Slightly overbend the tube and open it out to the correct angle to release the spring, then pull it out with the twine.

Using a pipe bender

Although you can hire bending springs to fit the larger pipes, it isn't easy to bend 22 or 28mm (¾ or 1in) tube over your knee – so it is well worth hiring a pipe bender to do the job.

Hold the pipe against the radiused former and insert the straight former to support it. Pull the levers towards each other to make the bend, and then open up the bender to remove the pipe.

Getting the bends in the right place

It is difficult to position two or more bends accurately along a single length of pipe. If you want to fit an alcove, for example, it's easier to bend individual lengths of pipe to fit each corner, then cut the tubes where they overlap and insert joints.

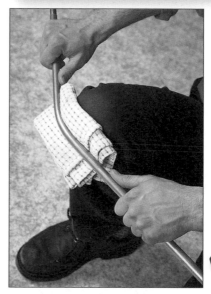

Plumbers' bending springs

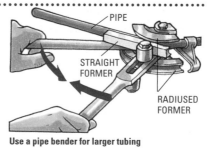

PIPE

STRAIGHT FORMER

RADIUSED FORMER

Use a pipe bender for larger tubing

CUT HERE   CUT HERE   CUT HERE

Bend separate lengths of pipe to fit an alcove

● **Annealing pipe**
When you are working with large-diameter copper pipe, play the flame of a gas torch around the area of the intended bend until the metal is cherry red, then allow it to cool. The pipe will bend with minimal effort, using a bending spring.

Using the spring
Bend the pipe against your padded knee. If you anneal the pipe pipe, (see above) be sure to allow it to cool before bending it.

Supporting pipe runs
Place a plastic or metal clip at 1m (3ft) intervals along a horizontal run of 15mm (½in) pipe. Increase the spacing to every 1.5m (4ft 6in) on a vertical run. In the case of larger pipes, increase the spacing a little more.

PLASTIC   METAL

☞ **SEE ALSO:** Connecting plastic to metal plumbing 25, External spring 76, Tube bender 76

Plastic plumbing

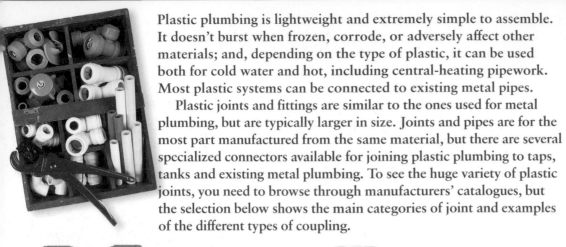

Plastic plumbing is lightweight and extremely simple to assemble. It doesn't burst when frozen, corrode, or adversely affect other materials; and, depending on the type of plastic, it can be used both for cold water and hot, including central-heating pipework. Most plastic systems can be connected to existing metal pipes.

Plastic joints and fittings are similar to the ones used for metal plumbing, but are typically larger in size. Joints and pipes are for the most part manufactured from the same material, but there are several specialized connectors available for joining plastic plumbing to taps, tanks and existing metal plumbing. To see the huge variety of plastic joints, you need to browse through manufacturers' catalogues, but the selection below shows the main categories of joint and examples of the different types of coupling.

Plastic supply pipes are made to the same standard sizes as metal pipework, but there may be a slight variation in wall thickness from one manufacturer's stock to another.

Straight connectors
For joining two pipes end to end.
1 For pipes of equal diameter – *push-fit*.

1

2

3

Elbows
For joining two pipes at an angle.
2 Elbow 45° – *solvent weld*.
3 Elbow 90° – *push-fit*.

Adaptors
To join dissimilar pipes.
4 Plastic-to-copper connector – *push-fit* and *compression joint*.

4

6

Tees
For joining three pipes.
5 Unequal tee for joining 15mm (½in) branch pipe to main pipe run – *push-fit*.

5

7

Fittings
Manufacturers supply pipe connectors and valves that can be attached to plastic pipes.

6 Tap connector with threaded nut for connecting supply pipe to tail of tap – *push-fit*.

7 Tank connector joins pipes to storage tanks and cisterns – *push-fit*.

8 Stopcock – *push-fit*.

8

Chlorinated polyvinyl chloride (cPVC)
A versatile plastic suitable for hot and cold supply. It can even withstand the temperatures that are required for central-heating systems.

Polybutylene (PB)
A tough, flexible plastic pipe used for hot and cold supply, and central heating. Available in standard lengths or continuous coils, PB resists bursting when frozen. It will sag if unsupported.

Cross–linked polyethylene (PEX)
Although it expands considerably when it is heated, PEX is used to make pipes that supply hot and cold water and for underfloor heating systems. However, it tends to sag, so is unsuitable for surface running. A PEX pipe resists bursting when subjected to frost. Twin-wall PEX, with an oxygen-diffusion barrier in the form of an aluminium layer sandwiched between the walls, is semi-rigid.

Medium–density polyethylene (MDPE)
This plastic is widely used for underground domestic supply pipes. The pipes, normally coloured blue, can be laid in continuous lengths and are resistant to pressure and corrosion.

Bending plastic pipes

● **Oxygen-diffusion barriers**
There's some concern that a small amount of oxygen drawn through the walls of plastic central-heating pipes contributes to the corrosion of the system. To prevent this happening, an oxygen-diffusion barrier is built into the walls of the pipe.

Flexible pipes can be bent cold to a minimum radius of eight times the pipe diameter. Use a pipe clip at each side of the bend to hold the curve, or use a special corner clamp. It is easy to thread flexible pipe around obstacles or run it under floorboards.

It's possible to bend a rigid plastic pipe by heating it gently. Pass the flame of a gas torch over the area that you want to bend. Keep the flame moving and revolve the pipe. When the pipe is soft enough, bend it by hand on a flat surface. Hold it still till the plastic hardens again. Wear thick leather gloves when handling hot plastic.

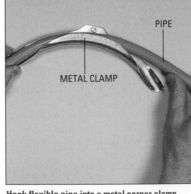

PIPE

METAL CLAMP

Hook flexible pipe into a metal corner clamp

MDPE

PEX

PB

cPVC

☞ **SEE ALSO:** Push-fit joints 25–6, Solvent-weld joints 26

Some plastic supply pipes can be connected using solvent-weld joints (as described for waste systems), but it is easier and more convenient to use the push-fit connectors shown below.

Push-fit joints

When the pipe is inserted, an O-ring seals in the water in the normal way and (depending on the model) a special plastic grab ring, or a collet with stainless-steel teeth, grips the tube securely to prevent water under mains pressure forcing the joint apart. Joints fitted with collets can be disconnected easily, but to dismantle the

Grab-ring push-fit joint
A grab ring holds the pipe, to resist water under pressure.

other type of push-fit joint, it's necessary to remove the retaining cap and prise open the grab ring, using a special tool.

Push-fit joints are more obtrusive than their solvent-welded equivalents – but the speed and simplicity with which you can assemble them more than compensates.

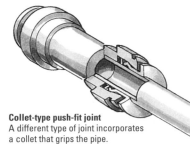

Collet-type push-fit joint
A different type of joint incorporates a collet that grips the pipe.

Special adaptor couplings are needed in order to connect most types of plastic pipe to copper or galvanized-steel plumbing. To join polybutylene pipe to copper, insert a metal support sleeve, then use a standard brass compression joint, or use a push-fit connector to join copper pipes to a polybutylene run. Cut and deburr the copper pipe carefully before pushing it into the joint.

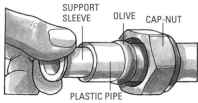

SUPPORT SLEEVE   OLIVE   CAP-NUT

PLASTIC PIPE

Joining plastic pipe with a compression fitting
Insert support sleeve before tightening the joint.

Collet-type joints

Push-fit joints that incorporate collets are particularly easy to assemble. Cut the end of the pipe square, push it into the socket until it comes up against the internal stop, then pull on the pipe to check that the joint is secure.

If you need to dismantle a joint, hold the collet in with your fingertips (1) and pull the pipe out of the socket.

Join metal pipes the same way, but remove burrs and sharp edges to prevent tearing the O-ring. Provide extra grip by slipping a collet clip into the grooved collar (2).

● **Supporting pipe runs**
Plastic pipework should be supported with clips or saddles similar to those used for metal pipe, but because it is more flexible you will have to space the clips closer together. Check with the manufacturers' literature to establish the exact dimensions. If you plan to surface-run flexible pipes, consider ducting or boxing-in because it's difficult to make a really neat installation.

Using grab-ring joints

Cut polybutylene pipe to length with the special shears that are supplied by the manufacturer (1) – or alternatively use a sharp craft knife. Provided that you make the cut reasonably square, the joint will be watertight.

Push a metal support sleeve into the pipe (2), and, if necessary, smear a little

silicone lubricant around the end of the pipe and inside the socket (3).

Push the prepared pipe firmly a full 25mm (1in) into the socket (4). As the joint can revolve freely around the pipe after connection without breaking the seal, there is no problem when aligning tees and elbows with other pipe runs.

 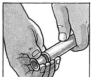

1 Cut pipe to length   **2 Insert metal sleeve**

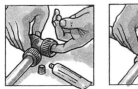

3 Apply lubricant   **4 Push pipe into joint**

Dismantling a joint

If you need to dismantle a joint to alter a system, unscrew the cap and pull out the pipe. Slide off the rubber O-ring, then prise off the grab ring, using a special demounting tool (see right). Never try to reuse a grab ring.

To reassemble the joint, insert the O-ring into the fitting, followed by the grab ring – with its slots facing outwards. Replace the retaining cap and hand-tighten it, ready to insert the pipe.

Push the pipe into the joint, using the technique described above. Never try to assemble the fitting like a compression joint, or it will blow out under pressure.

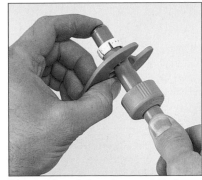

Prise open the grab ring, using a special tool

Repairing a weeping joint

A push-fit joint on a supply pipe may leak if the pipe is not pushed home fully, or if the O-ring is damaged.

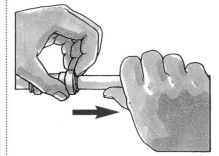

1 Hold the collet in with your fingertips

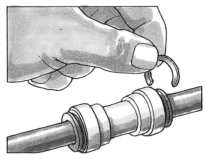

2 Slip collet clip into grooved collar

Cutting plastic pipe
Polybutylene pipe is easy to cut, using special shears.

☛ **SEE ALSO:** Supporting pipes 23, Solvent-weld joints 26, Hacksaws 74–5, Files 78

Plastic waste pipes

Plastics are complex materials, each having its own properties. Consequently, a technique or material that is suitable for joining one plastic may not be suitable for another.

To make watertight joints, it's vital to follow the manufacturer's instructions carefully, and to use the particular solvents and lubricants that are recommended. The examples on the right illustrate common methods for connecting plastic waste pipes and joints.

Types of plastic

Plumbing manufacturers have a wide variety of plastics to draw upon, each with its own special characteristics.

Modified unplasticized polyvinyl chloride (MuPVC)
A hard plastic, used for solvent-weld waste pipe and fittings. It is resistant to most domestic chemicals, and is not affected by ultra-violet light when used outdoors. It is slightly more flexible than uPVC, which is used for soil pipes with push-fit and solvent-weld joints.

Polypropylene (PP)
A slightly flexible plastic with a somewhat waxy feel, used for waste systems. It's impossible to glue PP, so it is assembled with push-fit joints.

Acrylonitrile butadiene styrene (ABS)
A very tough plastic that is equally suited to hot and cold waste. It can be either solvent-welded or compression-jointed.

Joints and fittings
As well as the usual types of joint, waste systems also include easy-flow swept bends and tees for efficient drainage.

Fittings
1 Bottle trap for sink or basin – *compression joint.*

Tees (T-joints)
2 Swept tee with access plug – *push fit.*
3 Branch 45°– *solvent-weld.*

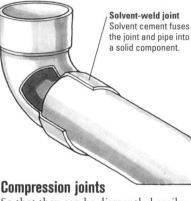

3

Bends and elbows
4 Elbow 90°– *push-fit.*

5 Bend 90°– *solvent-weld.*

5

PLASTIC WASTE-PIPE SIZES	
Overflow pipes	22mm (¾in)
Washbasin waste pipes	32mm (1¼in)
Bath/shower and sink waste pipes	40mm (1½in)
Soil pipe	110mm (4in)

Solvent-weld joints
Lengths of pipe are linked by simple socketed connectors. As they are assembled, solvent is introduced – which dissolves the surfaces of the mating components. As the solvent evaporates, the joints and pipes are literally fused together into one piece of plastic. Solvent-weld joints are sometimes used for supply pipes, but the technique is more commonly employed for waste systems.

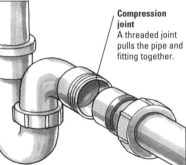

Solvent-weld joint
Solvent cement fuses the joint and pipe into a solid component.

Compression joints
So that they can be dismantled easily, sink, bath and washbasin traps are often connected to the pipework by means of compression joints that incorporate a rubber ring or washer to make the joint watertight.

Compression joint
A threaded joint pulls the pipe and fitting together.

Push-fit joints
Because a waste system is never under pressure, a pipe run can be constructed by simply pushing plain pipes into the sockets of the joints. A captive rubber seal in each socket holds the pipe in place and makes the joint watertight.

Push-fit joint
A rubber ring inside the sleeve grips the end of the pipe.

☞ SEE ALSO: Plastic supply pipes 24

It's important to follow the instructions supplied with any particular brand of pipe or fitting, but the methods given below and on the facing page describe the basic techniques for connecting plastic pipes.

Keep solvents away from children. Don't inhale solvent fumes, and never smoke when welding joints – fumes from some solvents become toxic if inhaled through a cigarette.

Work carefully and avoid spilling solvent cement – it will etch the surface of the pipe-work and damage some other plastics, as well.

Making push-fit joints

Cut the pipe to length and chamfer the end, as for solvent-weld joints. Wipe the inside of the socket with the recommended cleaner, and lubricate the pipe with a little of the silicone lubricant supplied with it.

Push the pipe into the joint right up to the stop, and mark the edge of the socket on the pipe with a pencil (1).

Withdraw the pipe about 9mm (⅜in) (2), to allow the pipe to expand when subjected to hot water.

1 Mark the edge of the socket on the pipe

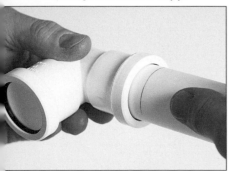

2 Withdraw the pipe about 9mm (⅜in)

Making solvent-weld joints

While the sequence of illustrations on the right shows large-diameter waste pipe, the methods described are equally valid for joining plastic supply pipe.

Cut the pipe to length with a fine-tooth saw, allowing for the depth of the joint socket. To make sure your cut is square, wind a piece of notepaper round the tube, aligning the wrapped edge as a guide (1). Revolve the pipe away from you as you cut it. Smooth the end with a file (2).

Welding the joint
Push the pipe into the socket to test the fit, then mark the end of the joint on the pipe with a pencil (3). This will act as a guide for applying the solvent. You need to key both the outside of the pipe and the inside of the socket with fine abrasive paper before using some solvents (check the manufacturer's instructions).

Before dismantling elbows and tees, scratch the pipe and joint with a knife (4), to help you align them correctly when you reassemble the components.

Use a clean rag to wipe the surface of the pipe and fitting with the recommended spirit cleaner. Paint solvent evenly onto both components (5), then immediately push home the socket. (Some manufacturers recommend that you twist the joint to spread the solvent.) Align the joint properly and leave it for 15 seconds.

The pipe is ready for use with cold water after an hour. But don't pass hot water through the system until at least four hours have elapsed (depending on the manufacturer's recommendations) or, preferably, longer.

Allowing for expansion
Plastic pipes expand when subjected to hot water. Generally this is only a problem over a straight run more than 3m (10ft) in length – but check the manufacturer's recommendations.

Incorporate an expansion coupling with a push-fit rubber seal at one end that allows the pipe to slide in and out without putting other joints under load. Lubricate the end of the pipe with silicone grease before you insert it into the coupling.

Repairing a weeping joint
If a joint leaks, leave it to dry out naturally. Then apply a little more of the solvent cement to the mouth of the socket, allowing it to flow into the joint by capillary action.

You would have to drain a supply pipe before you could make this repair.

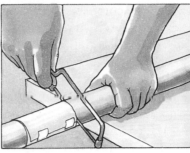

1 Use paper as a guide to keep the cut square

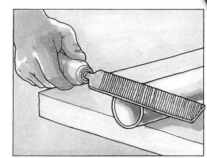

2 Smooth the end with a file

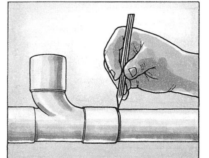

3 Assemble the joint and mark the socket

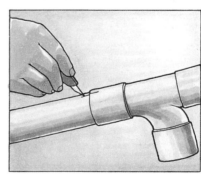

4 Scratch the pipe and joint to realign them

5 Paint solvent up to the pencil mark

● **Making compression joints to traps**
Traps with compression joints are made for connecting directly to a plain waste pipe (see opposite). Just slip the threaded nut onto the waste pipe, followed by the washer and then the rubber ring. Push the pipe into the socket of the trap and tighten the compression nut.

● **Repairing a weeping push-fit joint**
A push-fit joint will leak if the rubber seal has been pushed out of position. Dismantle the joint and check the condition of the seal.

☞ **SEE ALSO: Hacksaws 74–5, Files 78**

Replacing a WC suite

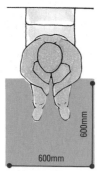

Replacing an old WC with a modern suite is a relatively straight-forward procedure, provided you can connect it to the existing branch of the soil pipe. However, if you are going to move a WC, or perhaps install a second one in another part of your home, you will have to connect to the main soil pipe itself or run the waste directly into the underground drainage system. In either case, it is worth hiring a professional plumber to make these connections.

Cisterns

From antique-style high-level cisterns to discreet close-coupled or concealed models, the choice is so wide that you're bound to find one to suit your requirements. Before buying, make sure the equipment carries the British Standard 'Kite mark' or complies with equivalent EC standards.

High-level cistern
If you simply want to replace an old-fashioned high-level cistern without having to modify the pipework, comparable cisterns are still available from plumbers' merchants.

Standard low-level cistern
Many people prefer a cistern mounted on the wall just above the WC pan. A short flush pipe from the base of the cistern connects to the flushing horn on the rear of the pan, while inlet and over-flow pipes can be fitted to either side of the cistern. Most low-level cisterns are manufactured from the same vitreous china as the WC pan.

Compact low-level cistern
Where space is limited, use a plastic cistern, which is only 114mm (4½in) from front to back.

Concealed cistern
A low-level cistern can be completely concealed behind panelling. The supply and overflow connections are identical to those of other types of cistern, but the flushing lever is mounted on the face of the panel. These plastic cisterns are utilitarian in character, with no concession to fashion or style, and are therefore relatively inexpensive. Don't forget that you will need to provide access for servicing.

Close-coupled cisterns
A close-coupled cistern is bolted directly to the pan, forming an integral unit. Both the inlet and overflow connections are made at the base of the cistern. An internal standpipe rises vertically from the overflow connection with the pan to protrude above the level of the water.

Space for a WC
You will need to allow a space at least 600mm (2ft) square in front of the pan.

600mm   600mm

600mm

600mm

WC pans

When visiting a showroom, you are confronted with many apparently different WC pans to choose from, but in fact there are two basic patterns – a washdown pan and a siphonic pan.

Siphonic pans
Siphonic pans need no heavy fall of water to cleanse them, and are much quieter as a result. A single-trap pan has a narrow outlet immediately after the bend, to slow down the flow of water from the pan. The body of water expels air from the outlet to promote the siphonic action. A double-trap pan is more sophisticated and exceptionally quiet. A vent pipe connects the space between two traps to the inlet that runs between the cistern and pan. As water flows along the inlet, it sucks air from the trap system through the vent pipe. A vacuum is formed between the traps, and atmospheric pressure forces the water in the pan into the soil pipe.

Washdown pans
Washdown pans work by simple displacement of waste by fresh water falling from the cistern. They are inherently more reliable than siphonic pans, but make considerably more noise when flushed.

Floor or wall exit?
When replacing a WC pan, check to see whether the new one needs to have a floor-exit or wall-exit trap.

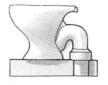

Floor-exit trap
S-traps are connected to a soil pipe that is then passed through the floor.

Wall-exit trap
The outlet from a P-trap connects to a soil-pipe branch located behind the pan.

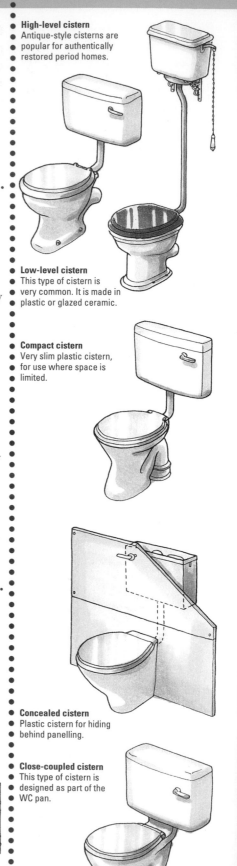

High-level cistern
- Antique-style cisterns are popular for authentically restored period homes.

Low-level cistern
- This type of cistern is very common. It is made in plastic or glazed ceramic.

Compact cistern
- Very slim plastic cistern, for use where space is limited.

Concealed cistern
- Plastic cistern for hiding behind panelling.

Close-coupled cistern
- This type of cistern is designed as part of the WC pan.

☞ SEE ALSO: **WC cisterns 12–14, Installing a WC suite 30, Bathroom planning 70**

Removing an old WC

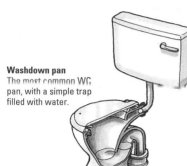

Washdown pan
The most common WC pan, with a simple trap filled with water.

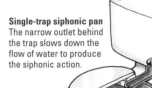

Single-trap siphonic pan
The narrow outlet behind the trap slows down the flow of water to produce the siphonic action.

Double-trap siphonic pan
Air is sucked out from between the two traps to create a vacuum.

Wall-hung pan
A wall-mounted pan, connected to a concealed cistern, leaves the floor clear for cleaning. Unless it is built into the masonry, the pan is supported by a metal bracket/stand.

Cut off the water supply, then flush the cistern to empty it. If you are merely renewing a cistern, you will have to disconnect the supply and overflow pipes with a wrench and loosen the large nut connecting the flush pipe to the base of the cistern. These connections are often corroded and painted – so it is easier to hacksaw through the pipes close to the connections if you intend to replace the entire suite.

Removing the old pan and cistern

Remove the fixing screws through the back of the cistern, or lift it off its support brackets and remove them. Lever the brackets off the wall with a crowbar if necessary.

Cut the overflow pipe from the wall with a cold chisel. Repair the plaster when you decorate the bathroom.

If the pan is screwed to a wooden floor, it will probably have a P-trap connected to a nearly horizontal branch soil pipe. Remove the pan's floor-fixing screws and scrape out the old putty around the pipe joint. Attempt to free the pan by pulling it towards you while

rocking it slightly from side to side.

If the joint is fixed firmly, smash the pan outlet just in front of the soil pipe with a club hammer (**1**). Protect your eyes with goggles. Stuff rags into the soil pipe to prevent debris falling into it, then chip out the remains of the pan outlet with a cold chisel (**2**). Work carefully, to preserve the soil pipe.

Smash an S-trap in the same way – and if the pan is cemented to a solid floor, drive a cold chisel under its base to break the seal. Chop out the broken fragments as before, and clean up the floor with a cold chisel.

1 Break the outlet of the pan with a hammer

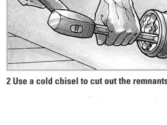

2 Use a cold chisel to cut out the remnants

Cutting the soil pipe

If you break the soil pipe while chipping out the pan outlet, cut the pipe square with a chain-link pipe cutter. To sever the pipe, clamp the chain of cutters around it, and work the tool's shaft back and forth. When you buy a push-fit pan connector (see below), make sure it is long enough to reach the severed pipe.

Pan to soil-pipe connection

Before you install the new suite, choose a push-fit flexible connector to join the pan to the soil pipe. There are connectors to suit most situations, even when the two elements are slightly misaligned. You may need an angled connector to join a modern horizontal-outlet pan to an old P-trap branch pipe (see opposite).

When selecting a connector, make a note of the following dimensions: the external diameter of the pan outlet, the internal diameter of the soil pipe, and the distance between the outlet and the pipe when the pan is installed.

Cutting a soil pipe
Use a chain-link cutter to cut a broken soil pipe square.

CUT HERE ➡

CUT HERE ➡

Removing an appliance
If fittings are corroded, remove the appliance by cutting through the flush pipe, overflow and pan outlet.

● **Lubricating connectors**
When installing plastic soil-pipe connectors, smear the surfaces lightly with a silicone lubricant.

PAN   CONNECTOR   SOIL PIPE

OFF-SET   ANGLED   BENT

Push-fit flexible pan connectors

☞ **SEE ALSO:** Turning off the water 6–9, WC cisterns 28, Chain-link cutter 75

Installing a new WC suite

Clean the floor and make good any damage before you begin to install a new WC suite.

Fitting and plumbing the suite

Push the plastic connector onto the pan outlet. Check that the inside of the soil pipe is clean and smooth, then slide the pan into place, pushing the connector firmly into the pipe.

Don't fix the pan yet. In a concrete floor, drill fixing holes and plug them. Level the pan on a bed of silicone sealant, using scraps of veneer or vinyl floorcovering as packing. Trim the packing flush when the job's complete.

Connect the flush pipe, then hold the cistern against the wall so you can mark fixing holes. Fix the cistern with non-corroding screws and washers, making sure it is level. You may have to use tap washers as packing behind the cistern to provide a clearance for the lid. Tighten the flush-pipe connection under the cistern.

Fit special protective sleeves into the pan-fixing holes and screw the pan to the floor, tightening the screws carefully in rotation to avoid cracking the pan. You can buy kits that provide all the necessary fixings for fitting WCs.

Run the new 15mm (½in) supply pipe to the float valve, fit a tap connector and tighten it with a wrench.

Attach a 22mm (¾in) overflow pipe, using the connector that's provided. Drill a hole through the nearest outside wall where an overflow is likely to be detected promptly. Slope the pipe a few degrees downwards, and let it project from the outer face of the wall at least 150mm (6in). If there isn't an external wall nearby, run the pipe to a combined waste and overflow unit on the bath. Alternatively, fit a tundish (see left) and run the overflow to the flush pipe or via a trap to a drain.

Turn on the water supply and adjust the float valve.

Tundish
A special funnel known as a tundish allows you to detect an overflow from a cistern.

Plumbing a WC
1 Overflow-pipe connector
2 22mm (¾in) overflow
3 Cistern
4 Float valve
5 Tap connector
6 15mm (½in) supply pipe
7 Flush-pipe connector
8 Flush pipe
9 Push-fit flexible connector
10 WC-pan outlet
11 Flexible outlet connector
12 Soil pipe

● **Fixing a new WC pan to the floor**
All manufacturers advise against the old-fashioned method of cementing a WC pan to a concrete floor. In fact, guarantees are usually invalidated if cement or a strong adhesive is used. If you can't screw the pan in place (see right), just rely on the bed of silicone sealant to bond the pan to the floor.

● **Installing a new high-level cistern**
A three-piece adjustable flush pipe allows you to hang a high-level cistern to one side of the pan. Fit a flow restrictor in the pan inlet if splashing water is a problem.

Typical pipe run
Red: Hot water
Blue: Cold water

SMALL-BORE WASTE SYSTEMS

The siting of a WC is normally limited by the need to use a conventional 110mm (4in) soil pipe and to provide sufficient fall to discharge the waste into the soil stack. By using an electrically driven pump and shredder unit, you can discharge WC waste through a 22mm (¾in) pipe up to 50m (55yd) away from the stack. The shredder will even pump vertically, to a maximum height of about 4m (12ft).

You can run the small-bore pipework through the narrow space between a floor and ceiling. Consequently, a WC can be installed as part of an en-suite bathroom, in a basement, even under the stairs, provided that the space is adequately ventilated.

The unit is designed to accept any conventional P-trap WC pan. It is activated by flushing the cistern, and switches off about 18 seconds later. It must be wired to a fused connection unit – via a suitable flex outlet if it is installed in a bathroom.

The waste pipe can be connected to the soil stack using any standard 32mm (1¼in) pipe boss, provided the manufacturer supplies a 22 to 32mm (¾ to 1¼in) adaptor. A WC waste pipe must be connected to the soil stack at least 200mm (8in) above or below any other waste connections.

Before you install a small-bore waste system, check that these systems are approved by your local water supplier.

Small-bore waste system for a WC
The shredding unit fits neatly behind a P-trap WC pan. When situated in a bathroom, the unit must be wired to a flex outlet. Otherwise, it can be connected directly to a fused connection unit.

☛ SEE ALSO: Adjusting float valves 8, Connecting pipes 19–27, Concealing pipework 31, Overflow 47, 81, Tank connector 49, Fused connection units 72, Float valves 81

Whether you're modifying existing plumbing or running pipework to a new location, fitting a washbasin in a bathroom or guest room is likely to present few difficulties provided you give some thought to how you will run the waste to the vertical stack. The waste pipe must have a minimum fall or slope of 6mm (¼in) for every 300mm (1ft) of pipe run and should not be more than 3m (10ft) long.

Selecting a washbasin

Wall-hung and pedestal washbasins are invariably made from vitreous china, but basins that are supported all round by a counter top are also available in pressed steel and plastic.

Select the taps at the same time, to ensure that the basin of your choice has holes at the required spacing to receive the taps – or no holes at all if the taps are to be wall-mounted.

Pedestal basins

The hollow pedestal provides some support for the basin and it conceals the unsightly supply and waste pipes.

Wall-hung basins

Older wall-hung basins are supported on large screw-fixed brackets, but a modern concealed mounting is just as strong provided the wall fixings are secure. Check that you can screw into the studs of a timber-frame wall or hack off the lath-and-plaster and install a mounting board. If you want to hide pipes, consider some form of panelling.

Corner basins

Handbasins that fit into the corner of a room are space-saving, and the pipework can be run conveniently through adjacent walls or concealed by boxing them in across the corner.

Recessed basins

In a cloakroom or WC where space is very limited, a small handbasin can be recessed into one of the walls. Also, you can recess a standard basin to conceal the plumbing.

Counter-top basins

In a large bathroom or bedroom, you can fit a washbasin or pair of basins into a counter top as part of a built-in vanity unit. Cupboards below provide ample storage for towels and toiletries, while also hiding the plumbing.

Pedestal basin

Wall-hung basin

Corner basin

Recessed basin

Counter-top basin

With carefully designed pipe runs, it should be possible to plumb your house without a single pipe being visible. In practice, however, there are always situations where you have no option but to surface-run some pipes.

You can minimize the effect by taking care to group pipes together neatly and keeping runs both straight and parallel. When painted to match the skirtings or walls, such pipes are barely visible.

Alternatively, using softwood battens and plywood, you can make your own accessible ducting to bridge the corner of a room; or construct a false skirting that is deep enough to contain the pipes.

For total accessibility, you can use proprietary ducting made from PVC. This is manufactured in a range of sizes, to contain grouped or individual pipes.

Clip pipes to the wall

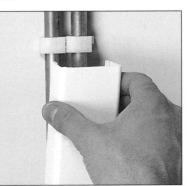
Snap on the cover-strip

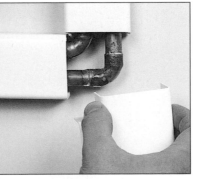
Snap on corner covers

Space for a basin
Allow extra elbow room for washing hair – a space 1100mm (3ft 8in) x 700mm (2ft 4in) should be sufficient.
To suit most people, position the rim of a basin 800mm (2ft 8in) from the floor.

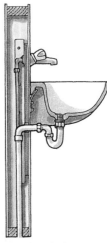

Mounting a basin
Fix a wall-mounted basin and taps to an exterior-grade plywood board fixed to a stud partition.

Hide your pipes inside plastic ducting

☞ **SEE ALSO:** Fitting a washbasin 33, Bathroom planning 70

Selecting taps

Taps – which are now very much a fashion item – come in different styles and colours. Not all taps are built to last, so check the quality if you are buying for the long term. Chromium-plated brass taps are the most durable. Check that the taps you are considering will fit the layout of holes in the basin for which they're intended.

● **The right pressure**
Some taps imported from the Continent have relatively small inlets and are intended for use with mains-pressure supply only. These taps will not work efficiently if they are connected to a low-pressure tank-fed supply.

Types of tap

The majority of washbasins are fitted with individual taps for hot and cold water. While capstan-head taps are still manufactured for use in period-style bathrooms, most modern taps have a shrouded head made of metal or plastic.

A lever-head tap turns the water from off to full on with one quarter turn only. This type is convenient for the elderly or disabled, who may have difficulty in manipulating other taps.

In a mixer tap, hot and water cold are directed to a common spout. Water is supplied at the desired temperature by adjustment of the two valves. With a single-lever mixer tap, flow rate and temperature are controlled by adjusting the one lever.

Washbasin mixer taps sometimes incorporate a pop-up waste plug. A series of interlinked rods, operated by a button or small knob on the centre of the mixer, open and close the waste plug in the basin.

Normally, the body of the tap (which connects the valves and spout) rests on the upper surface of the washbasin. But it is also possible to mount it in its entirety on the wall above the basin. Another alternative is for the valves to be mounted on the basin and divert hot and cold water to a spout mounted on the wall above.

Single-lever mixer tap
Moving the lever up and down turns the water on and off. Swinging it from one side to the other gradually increases the temperature, by mixing more hot water with the cold.

Tap mechanisms

Over recent years there have been some revolutionary changes in the design of taps that have made them easier to operate and simpler to maintain.

Rising-spindle taps
This traditional tap design has a washer on the end of a spindle that rises as the tap is turned on. It is a simple, rugged mechanism that lasts for years.

Non-rising-spindle taps
Theoretically, these taps should exhibit fewer problems than rising-spindle taps, because the mechanism imposes less wear on the washer. In practice, however, the spindle's fine thread is prone to wear, and there is potential for misalignment caused by the circlip that holds the mechanism in place.

Ceramic-disc taps
With these taps, precision-ground ceramic discs are used in place of the traditional rubber washer. One disc is fixed and the other rotates until the waterways through them align and water flows. There is minimal wear, as hard-water scale or other debris is unlikely to interfere with the close fit of the discs. However, if a problem does develop, the entire inner cartridge and the lower seal can be replaced.

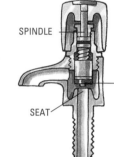

Rising-spindle tap
Traditional taps are made with a rising spindle.

RISING SPINDLE
WASHER
SEAT

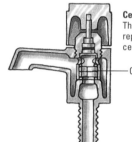

Non-rising-head tap
A spindle that doesn't revolve reduces wear on the washer.

SPINDLE
WASHER
SEAT

Ceramic-disc tap
The rubber washer is replaced with rotating ceramic discs.

CERAMIC DISCS

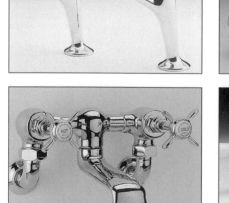

Basin and bath taps
(top row – left to right)
Single capstan-head pillar taps
Single-lever taps
One-hole basin mixer
(bottom row – left to right)
Two-hole bath mixer
Three-hole basin mixer
Shower-mixer deck

☞ **SEE ALSO: Repairing taps 33**

REMOVING OLD TAPS

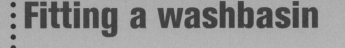

Fitting a washbasin

When replacing taps, you will want to use the existing plumbing if possible, but disconnecting old, corroded fittings can be difficult.

Apply some penetrating oil to the tap connectors and to the back-nuts that clamp the tap to the basin. While the oil takes effect, shut off the cold and hot water supply to the taps.

If necessary, apply heat with a gas torch to break down the corrosion – but wrap a wet cloth around nearby soldered joints, or you may melt the solder. Take care that you do not damage plastic fittings and pipes, and protect flammable surfaces with a ceramic tile. Try not to play the flame onto a ceramic basin.

A cranked spanner fits basin and bath taps

Cranked spanners

It is not always possible to engage the nuts with a standard wrench. Instead, hire a special cranked spanner designed to reach into the confined spaces below a basin or bath. You can apply extra leverage to the spanner by slipping a stout metal bar or wrench handle into the other end.

Removing a stuck tap

Even when you have disconnected the pipework and back-nut, you may find that the taps are stuck in place with putty. Break the seal by striking the tap tails lightly with a wooden mallet. Clean the remnants of putty from around the holes in the basin, then fit new taps. If the tap tails are shorter than the originals, buy special adaptors designed to take up the gaps.

Releasing a tap connector
Use a special cranked spanner to release the fixing nut of a tap connector.

Turn off the supply of water to an old basin before you disconnect it.

Removing an old basin

If you want to use existing plumbing, loosen the compression nuts on the tap tails (see left) and trap. Otherwise, cut through the waste and supply pipes at the point where you can most easily connect new plumbing **(1)**.

Remove any fixings holding the basin to its support brackets or pedestal, and lift it from the wall. Apply penetrating oil to the brackets' wall fixings, in the hope that you'll be able to remove them without damaging the plaster – but as a last resort, lever the brackets off the wall. Take care not to break cast-iron fittings, as they can be quite valuable.

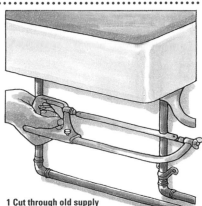

1 Cut through old supply pipes with a hacksaw

Fitting new taps

Fit new taps to the basin before you fix it to the wall. Slip the plastic washer supplied with the tap onto its tail, then pass the tail through the hole in the basin. (If no washer is supplied, spread some silicone sealant around the top of the tail and beneath the base of the tap.)

With the basin resting on its rim, slip a second washer onto the tail then hand-tighten the back-nut to clamp the tap onto the basin **(2)**. Check that the spout faces into the basin, then tighten the back-nut carefully with a cranked spanner (see left).

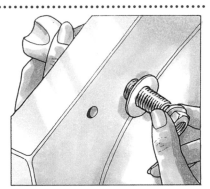

2 Slip the back-nut onto the tail of the tap

Fixing the basin to the wall

Get an assistant to hold a wall-hung basin against the wall at the required height while you use a spirit level to check that it is horizontal. Mark the fixing holes for the wall bracket **(3)**.

For a pedestal basin (see right), place the pedestal in position, then sit the basin on it and mark the fixing holes. Lay the basin (and pedestal) to one side while you drill and plug the holes **(4)**.

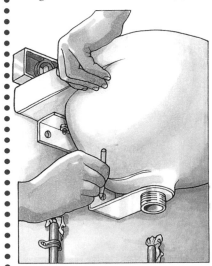

3 Mark the fixing holes on the wall

4 Drill and plug the holes

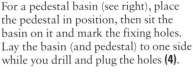

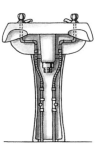

Pedestal basins
Run pipework up to and behind a pedestal. Fix the basin to the wall with screws. Some basins are attached to the pedestal with clips, or may need bonding to it with silicone sealant. Screw the pedestal to the floor.

☞ **SEE ALSO:** Turning off the water 6, Connecting pipes 19–27, Gas torch 21, 77, Hacksaws 74–5, Spanners and wrenches 77–8

Connecting a basin

Once you have fitted the new taps and mounted the basin securely to the wall, complete the installation by connecting the trap and waste pipe, followed by the supply pipes for hot and cold water. Fit isolating valves to the supply pipes, to make servicing easier in the future. If you are installing a pedestal basin, fit the trap before fixing the basin to the wall.

Pressed-metal basin
When you fit taps to a pressed-metal basin, slip built-up 'top-hat' washers onto the tails to cover the shanks. The basin itself may be supplied with a rubber strip to seal the joint with the counter top. It will need a combined waste and overflow, like a bath.

● **Counter-top basin**
Manufacturers supply a template for cutting the hole in the counter top to receive the basin. Run mastic around the edge to seal a ceramic basin, and clamp it with the fixings supplied.

Bottle trap
It is easy to remove a blockage from a bottle trap, because the entire base of the trap can be unscrewed by hand.

Plumbing a washbasin
1 Tap back-nut and washer
2 Flexible copper pipe
3 15mm (½in) supply pipe
4 Isolating valves
5 Waste outlet (slot faces overflow)
6 Waste back-nut and washer
7 Bottle trap
8 32mm (1¼in) waste pipe

Typical pipe runs
Red: Hot water
Blue: Cold water

Fitting trap and waste

Fit the waste outlet into the bottom of the basin as described for taps, using washers or a silicone sealant to form a watertight seal. The basin will probably have an integral overflow running to the waste, in which case ensure that the slot in the waste outlet aligns with the overflow. Tighten the back-nut under the basin, while holding the outlet still by gripping its grille with pliers.

If you can use the existing waste pipe, connect the trap to the waste outlet and to the end of the pipe. A two-part trap provides some adjustment for aligning with the old waste pipe.

To run a new 32mm (1¼in) waste pipe, cut a hole through the wall with a masonry core drill. Run the pipe, with sufficient fall – 6mm (¼in) per 300mm (1ft) run – to terminate over the hopper on top of the outside downpipe or feed into a soil pipe (see far right). Fix the waste pipe to the wall with saddle clips.

Connecting the taps

You can run standard 15mm (½in) copper or plastic pipes to the taps and join them with tap connectors, but it is easier to use short lengths of flexible corrugated copper pipe designed specially for tap connection. They can be bent by hand to allow for any slight misalignment between the supply pipes and tap tails, and they are easy to fit behind a pedestal. Each pipe has a tap connector at one end and a capillary or compression joint at the other.

Connect the corrugated pipes to the tap tails, leaving them hand-tight only. Then run new branch pipework to meet the corrugated pipes, or connect them to the existing plumbing. Make soldered or compression joints to connect the pipes. Use a cranked spanner to tighten the tap connectors. Turn on the water supply and check the pipes for leaks; if you need to repair a weeping soldered joint, drain the system.

A proprietary pipe boss is used to connect a basin waste pipe to a single-stack plastic soil pipe. There are various ways of connecting the boss, one of the simplest being to clamp it with a strap.

Mark where the basin waste meets the soil pipe, and use a hole saw to cut a hole of the recommended diameter **(1)**. Smooth the edge of the hole with abrasive paper.

Wipe both contacting surfaces with the manufacturer's cleaner, then apply gap-filling solvent cement around the hole. Strap the boss over the hole and tighten the bolt **(2)**.

Insert the rubber lining in the boss, in preparation for the waste pipe **(3)**.

Lubricate the end of the pipe and push it firmly into the boss **(4)**. Clip the pipe to the wall.

1 Cut a hole in the pipe with a hole saw

2 Strap the boss over the hole

3 Insert the rubber lining

4 Push the waste pipe into the boss

☞ **SEE ALSO:** Draining the system 8, Connecting pipes 19–27, Cutting soil pipes 29, Fitting taps 33, Mounting a basin 33, Cranked spanner 77

Choosing a new bath

An antique cast-iron bath can be worth a great deal of money, so get a quotation from a dealer if you decide to replace it. Bear in mind that there are companies that re-enamel old baths, and some will even spray them in your bathroom. However, if your old bath has deteriorated badly, it may prove more economical to replace it – and a cracked bath will be completely beyond repair.

Selecting a bath

You can purchase reproduction or even restored Victorian baths in cast iron from specialist suppliers, but they are likely to be expensive. In practical terms, a cast-iron bath is far too heavy for one person to handle – even two people would have difficulty carrying one to an upstairs bathroom. Also, while a cast-iron bath can look splendid when left freestanding in a room, it may be virtually impossible to clean behind it, and panelling-in the curved and often tapering shape is rarely successful.

Nowadays, the majority of baths are made from enamelled pressed steel, acrylic or glass-reinforced plastic. Two people can handle a steel bath with ease, and you could carry a plastic bath on your own. Although modern plastic baths are strong and durable, some are harmed by abrasive cleaners, bleach and especially heat. It is not advisable to use a gas torch near a plastic bath.

So far as style and colour are concerned, there's no lack of choice in any material, although the more unusual baths are likely to be made of plastic. Nearly every bath comes with matching panels, and optional features such as hand grips and dropped sides to make it easier to step in and out. Taps do not have to be mounted at the foot of

the bath. Many manufacturers offer alternative corner- or side-mounting facilities, and some will even cut tap holes to your specification.

You can order bath tubs that double as a jacuzzi – but the plumbing is somewhat complicated, so you will need to have them professionally installed.

Rectangular bath

A standard rectangular bath is still the most popular and economical design. Baths vary in size from 1.5 to 1.8m (5 to 6ft) in length, with a choice of widths from 700 to 800mm (2ft 4in to 2ft 8in).

Corner bath

A corner bath actually occupies more floor area than a rectangular bath of the same capacity, but because the tub is turned at an angle to the room it may take up less wall space. By virtue of its design, a corner bath usually provides some shelf space for essential toiletries.

Round bath

A round bath is likely to be impractical in most bathrooms – but if you are converting a spare bedroom, you may decide to make the bath a feature of the interior design as well as a practical appliance.

RENOVATING BATH ENAMEL

You can buy two-part paints prepared specifically for restoring the enamel surface of an old bath, sink or basin.

To achieve a first-class result, the bath must be scrupulously clean and dry – so tape plastic bags over the taps to prevent water dripping into the bath, and work in a warm atmosphere where condensation will not occur. To remove any grease, wipe the surface with a cloth dampened with white spirit; then paint the bath from the bottom upwards, in a circular direction. This type of paint is self-levelling, so don't brush it out too much. Pick up runs immediately, and work quickly to keep wet edges fresh.

For a professional finish, hire a company that will send an operator to spray the bath *in situ*. The process shouldn't take longer than two or three hours. First, the bath is cleaned chemically; then a grinder is used to key the surface and remove heavy stains. At the same time, chipped enamel can be repaired. Finally, surrounding areas are masked before the bath is sprayed.

Access to a bath
Allow a 1100 x 700mm (3ft 8in x 2ft 4in) space beside a bath so that it's possible to climb in and out safely, and for bathing younger members of the family.

Restoring an enamel surface
Use a two-part paint system to restore the enamel surface of an old bath.

Supporting a plastic bath

A frame with adjustable feet is supplied to cradle a flexible plastic bath. The parts need to be assembled before the bath is fitted into place.

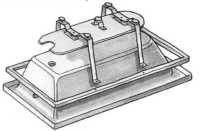

Assembling the cradle
Turn a bath onto its rim to fit the cradle.

● **Selecting taps for a bath**
In design and style, bath taps are identical to basin taps; but they are proportionally larger, with 22mm (¾in) tails. Some bath mixers are designed to supply water to a sprayhead, either mounted telephone-style on the mixer itself or hung from a bracket mounted on a wall above the bath.

☛ **SEE ALSO: Selecting taps 32, Plumbing a bath 36, Shower mixers 38**

Plumbing a bath

Once a bath is fitted close to the wall, it can be difficult to make the joints and connections – so fit the taps, overflow and trap before you push the new bath into position (see bottom right). Set the adjustable feet to raise the rim of the bath to the required height, and check it for level along its length and width. If the bath has small feet, cut two boards to go under them to spread the point load over a wider area.

Fitting the taps

Fit individual hot and cold taps as for a washbasin. Fitting a mixer tap is a similar procedure, but most mixers are supplied with a long sealing gasket that slips over both tails. Lower the tails through the holes in the rim, then slip top-hat washers onto them and tighten both back-nuts to clamp the mixer securely to the bath. Fit a flexible 22mm (¾in) copper pipe (similar to those used for washbasin taps) onto each tail.

These flexible pipes allow for the easy adjustment that will be necessary if the joints are slightly misaligned. Alternatively, attach short lengths of standard 22mm (¾in) copper or plastic pipe with tap connectors, in preparation for jointing to the pipe run.

Waste/overflow units
A flexible tube takes any overflow water to the trap.

Compression unit
Runs to the cleaning eye on the trap.

Banjo unit
Slips over the tail of the waste outlet.

WC and bath overflow
Overflow from a WC joins the bath unit.

Shallow-seal trap
Use this type of trap when space is limited. It must discharge to a yard gully or hopper, not to a soil stack.

Plumbing a bath
1 Mixer tap
2 Mixer-tap gasket
3 Mixer back-nut and washer
4 Flexible copper pipe
5 Overflow unit
6 Waste outlet
7 Waste back-nut and washer
8 Deep-seal trap to 40mm (1½in) waste pipe
9 Supply pipes – 22mm (¾in)

Fitting waste and overflow

Fit a combined waste and overflow unit to the bath. A flexible plastic hose takes water from the overflow outlet at the foot of the bath to the waste outlet or trap. If you use a 'banjo' unit, you must fit the overflow before the trap; but the flexible pipe of a compression-fitting unit connects to the trap itself (see left).

Spread a layer of silicone sealant under the rim of the waste outlet, or fit a circular rubber seal. Before inserting its tail into the hole in the bottom of the bath, seal the thread with PTFE tape. On the underside, add a plastic washer; then tighten the large back-nut, bedding

the outlet down onto the sealant or the rubber seal. Wipe off excess sealant.

Connect the bath trap (see left) to the tail of the waste outlet with its own compression nut. (Fit a banjo overflow unit at the same time.)

Pass the threaded boss of the over-flow hose through the hole at the foot of the bath. Slip a washer seal over the boss, then use a pair of pliers to screw the overflow outlet grille on.

If you're using a compression-fitting overflow, connect the nut located on the other end of the hose to the cleaning eye of the trap.

Turn off the water supply before you drain the system.

Removing an old bath

Have a shallow bowl ready to catch any trapped water, then use a hacksaw to cut through the old pipes. The over-flow pipe from an old bath will almost certainly exit through the wall, so saw through the overflow at the same time.

If the bath has adjustable feet, lower them and then push down on the bath to break the mastic seal between the bathroom walls and the rim. Pull the bath away from the walls.

If a cast-iron bath is beyond restoration and therefore worthless, it is easier to break it up in the bathroom and carry it out in pieces. Drape a dust sheet over the bath; then, wearing gloves, goggles and ear protectors, smash it with a heavy hammer.

Hack the old overflow from the wall with a cold chisel, then fill the hole with mortar and repair the plasterwork.

Installing a new bath

Either run new 22mm (¾in) supply pipes or attach spurs to the existing ones, ready for connection to the flexible pipes already fitted on the bath taps.

Slide your new bath into position and adjust the height of the feet with a spanner. Use a spirit level to check that the rim is horizontal.

Adjust the flexible tap pipes and join them to the supply pipes. Connect a 40mm (1½in) waste pipe to the trap and run it to the external hopper or soil stack, as for a washbasin. Before fixing the bath panels, restore the water supply and check for leaks.

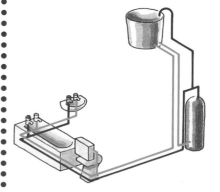
Typical tank-fed bathroom pipe runs
Red: Hot water. **Blue**: Cold water.

☞ SEE ALSO: **Draining the system 8, Connecting pipes 19–27, Fitting taps 33, Stack connection 34**

Choosing a shower

All showers, except for the most powerful, use less water than required for filling a bath. And because showering is generally quicker than taking a bath, it helps to alleviate the morning queue for the bathroom. For even greater convenience, install a second shower somewhere else in the house – this is one of those improvements that really does add value to your home.

Improvements in technology have made available a variety of powerful, controllable showers. However, many appliances are superficially similar in appearance, so it's important to read the manufacturers' literature carefully before you opt for a particular model.

Pressure and flow

When choosing a shower, it should be borne in mind that pressure and flow are not the same thing. For example, an instantaneous electric shower delivers water at high mains pressure, but a relatively low flow rate is necessary to allow the water to heat up as it passes through the shower unit.

A conventional gravity-fed supply system delivers hot water from a storage cylinder under comparatively low pressure, but often has a fairly high flow rate when measured in litres per minute. Adding a pump to this type of system can increase the pressure and flow rate. It is then possible to alter the flow and pressure ratio by fitting an adjustable showerhead that provides a choice of spray patterns, from needle jets to a gentle cascade (often called 'champagne').

This showerhead provides a choice of spray patterns

Gravity-fed showers

In many homes cold water is stored in a tank, from which it is fed to a hot-water cylinder situated at a lower level. Both the hot-water and cold-water pressures are determined by the height (known as the 'head') of this cold-water storage tank above the shower. Provided there is at least one metre (3ft) between the bottom of the tank and the showerhead, you should have reasonable flow rate and pressure.

If flow and pressure are insufficient for a satisfactory shower, it may be possible to improve the situation either by raising the tank or by installing a pump in the system.

Mains-pressure showers

You can supply some types of shower directly from the mains. In fact, one of the simplest to install is an instantaneous electric shower, which is designed for use with mains pressure.

Another alternative is to install a thermal-store cylinder. Mains-pressure water passes through a rapid heat exchanger inside the cylinder (see right). Yet another option is to store hot water in an unvented cylinder – which will supply high-pressure water to a shower without the need for a booster pump.

Nowadays showers are often supplied from combination boilers, though these often need to run at full flow to keep the boiler firing properly. Before buying a shower, check with the manufacturer of your boiler to ascertain whether there's likely to be a problem.

Drainage

Draining the used water away from a shower can be more of a problem than running the supply.

If it is not possible to run the waste pipe between the floor joists or along a wall, then you may have to consider relocating the shower. In some situations it may be necessary to raise the shower tray on a plinth in order to gain enough height for the waste pipe to fall (slope) towards the drain. Another way to overcome the problem is to install a special pump to take the waste water away from the shower.

Shower traps

When running the waste pipe to an outside hopper, you can fit a conventional trap – but these are relatively large, which can make for difficulties when installing the shower tray.

You could cut a hole in the floor, or substitute either a smaller, shallow-seal trap or a compact trap that includes a removable grid and dip tube for easy cleaning. Another possibility is to fit a running trap in the waste pipe at a convenient location, or install a self-sealing valve in the pipe.

A shower trap that is connected to a soil stack must have a water seal not less than 50mm (2in) deep. The easiest solution is to fit a compact trap, which is shallow enough to fit under most modern shower trays, but is designed to provide the necessary water seal. Or you could fit either a running trap or a self-sealing valve, as mentioned above.

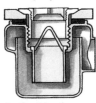

Thermal-store cylinder
Mains-fed water passes through a rapid heat exchanger on its way to the shower.
1 Mains feed
2 To shower
3 Other outlets
4 Boiler connections

Shower enclosures
If space permits, choose an enclosed shower cubicle (far left). However, there are a number of screens and plumbing options, which make an over-the-bath shower almost as efficient.

Running trap

Section through a compact shower trap

Cleaning compact traps
Compact traps for showers have a lift-out dip tube for easy cleaning.

☞ **SEE ALSO:** Thermal-store cylinders 37, 51, Booster pumps 39, 42, Unvented cylinders 51

Shower mixers

Installing an independent shower cubicle with its own supply and waste systems requires some prior experience of plumbing – but if you use an existing bath as a shower tray, then fitting a shower unit can involve little more than replacing the taps.

Bath/shower mixers

This type of shower is the simplest to install. It is connected to the existing 22mm (¾in) hot and cold pipes in the same way as a standard bath mixer, and the bath's waste system takes care of the drainage. Once you have obtained the right temperature at the spout by adjusting the hot and cold valves, you lift a button on the mixer to divert the water to the sprayhead via a flexible hose. The sprayhead can be hung from a wall-mounted bracket to provide a conventional shower, or hand-held for washing hair. The main disadvantage with this type of shower is that the controls are uncomfortably low to reach.

Since the supply pipes are already part of the bathroom's plumbing network, it's impossible to guard against fluctuating pressure unless the mixer is fitted with a thermostatic valve or you install a pressure-equalizing valve in the pipework. If the pressure is insufficient, fit a booster pump.

Don't fit a bath/shower mixer unless both the hot and cold water is under the same pressure, either high or low.

Bath/shower mixer
Fit this type of shower unit like an ordinary bath mixer.

Thermostatic mixer
This unit prevents excessive fluctuations in water temperature.

Manual shower mixers

A manual shower mixer can be fixed to the wall above a bath or situated in a separate shower cubicle. Manual mixers require their own independent hot and cold supply.

Simple versions are available with individual hot and cold valves, but most manual shower mixers have a single control that regulates the flow and temperature of the water. Single-lever ceramic-disc mixers operate exceptionally smoothly and, having few moving parts, are not so prone to hard-water scaling.

You can choose a surface-mounted unit or a nearly flush mixer with the pipework, connections and shower mechanism all concealed in the wall.

Thermostatic mixers

A thermostatic shower mixer is similar in design to a manual mixer but it has an extra control incorporated, to preset the water temperature. If the flow rate drops on either the hot or cold supply, a thermostatic valve rapidly compensates by reducing the flow on the other side. This is primarily a safety measure, to prevent the shower user being scalded should someone run a cold tap elsewhere in the house. Consequently, you can supply a thermostatic shower by means of branch pipes from the bathroom plumbing – but try to join them as near as possible to the cold tank and hot cylinder. The mixer can't raise the pressure of the supply, so you still need a booster pump if the pressure is low.

Thermostatic mixer mechanisms are usually based on wax-filled cartridges or bimetallic strips. Brand-new thermostatic valves respond extremely quickly to changes of temperature, but you can expect the rate to slow down as scale gradually builds up inside the mixer. Even when new, reaction time will be slower if the mixer is expected to cope with exceptionally hot water (above 65°C/149°F). At such high temperatures the hot-water ports are almost fully closed and the cold-water ones almost wide open, so there is very little margin for further adjustment.

The majority of thermostatic mixers can be used with the existing gravity-fed hot and cold supply, but it may be necessary to fit a booster pump. Check the manufacturer's literature carefully – since some showers perform well at low pressures, while others will be less than satisfactory.

Single-lever mixer
With this type of mixer, a single control is used to regulate flow and temperature.

An instantaneous electric shower is designed specifically for connection to the mains water supply, using a single 15mm (½in) branch pipe from the rising main. A non-return valve must be fitted close to the unit.

You can install an instantaneous shower practically anywhere, so long as drainage is feasible.

Incoming water is heated within the unit, so there is no separate hot-water supply to balance. The shower is thermostatically controlled to prevent fluctuations in pressure affecting the water temperature – in fact, it switches off completely if there is a serious failure of pressure. You can even buy an instantaneous shower with a shut-down facility: when you switch off, the water continues to flow for a little while to flush any hot water out of the pipework. This ensures that someone stepping into the cubicle immediately after another user isn't subjected to an unexpectedly hot start to their shower.

The electrical circuit

An instantaneous shower requires its own circuit from the consumer unit. A ceiling-mounted double-pole switch is connected to the circuit to turn the appliance on and off.

Surface-mounted or concealed

With most instantaneous showers, all plumbing and electrical connections are contained in a single mixer cabinet that is mounted in the shower cubicle or over the bath. However, you can buy showers with a slim flush-fitting control panel that is connected to a power pack installed out of sight – for example, under the bath behind a screw-fixed panel.

Fit a stopcock or miniature isolating valve in the supply pipe to allow the shower to be serviced.

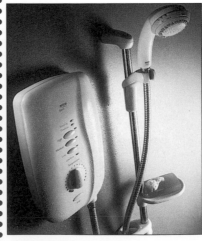

☞ **SEE ALSO:** Wiring a shower 71

Pump-assisted showers

High-performance showers have propagated a new generation of sprayheads, which offer a variety of spray patterns.

If you're thinking of upgrading an existing shower by installing an electric pump, it's worth finding out whether you can also substitute an adjustable sprayhead.

In addition to the standard shower spray, a simple adjustment is all that is needed to produce an invigorating jet to wake you up in the morning or a soft bubbly stream that is ideal for small children. Some sprayheads can also be adjusted to deliver a very light spray while you soap yourself or apply shampoo.

Cleaning a sprayhead

Gradually accumulation of lime scale blocks the holes in the sprayhead, and eventually this affects the performance of your shower. It's therefore essential to clean the sprayhead, the frequency of cleaning depending on the hardness of the water in the area where you live.

Remove the entire sprayhead from its hose or unscrew the perforated plate from the showerhead. Leave the sprayhead or plate to soak in a proprietary descalant until the scale has dissolved, then rinse thoroughly under running cold water.

Before you reattach the sprayhead or plate, turn on the shower to flush any loose scale deposits from the pipework.

Electrical installations

Electrical installations in a bathroom are potentially dangerous – which is why they must conform to the current Wiring Regulations compiled by the Institution of Electrical Engineers. Before you undertake the work, read the electrical section in this book and check the manufacturers' instructions carefully to make sure you understand the requirements for wiring in a bathroom. If you are in any doubt as to the procedure, or have not had previous experience, hire a qualified electrician.

Power showers

The pump-assisted 'power' shower is perhaps most people's concept of the ideal shower. The pump delivers water at a constant pressure and flow rate, eliminating the need for the minimum pressure normally required for a gravity-fed shower. Most power showers need a head of about 75 to 225mm (3 to 9in) to activate the pump when the mixer control is turned on. A pump can be used to boost the pressure and flow rate of stored hot and cold water, but not mains-fed water.

Ideally, the cold supply should be taken directly from the storage tank – not from branch pipes that feed other taps and appliances. The hot-water supply can be connected to the cylinder by means of a Surrey or Essex flange; this helps eliminate the tendency for the pump to suck in air from the vent pipe.

If the water is heated by an electric immersion heater, make sure the cylinder is fed by a dedicated cold feed and that the cold-feed gate valve is fully open. This is to prevent the top of the cylinder running dry and perhaps burning out the heater. If the cylinder is heated from a boiler, make sure the water temperature is controlled by a thermostat. If the water is too hot, the shower could splutter.

Power showers are frequently manufactured with an electrically driven pump built into the mixer cabinet that is mounted in the shower cubicle.

However, some pumps are designed for remote installation, with hot and cold pipes running to the pump then out again to the shower mixer. These freestanding pumps can also be used to improve the performance of an existing installation. The usual location for this type of pump is next to the hot-water cylinder in an airing cupboard – as low as possible, so that the pump remains full of water. However, there are also pumps that are designed to perform satisfactorily when mounted at a high level – even in the loft, if that is the only option available. In such situations, a single-impeller pump is best.

● **Water Regulations**
If the shower is mounted in such a way that the sprayhead could dangle below the rim of the bath or shower tray, you have to fit double-seal non-return valves in the supply pipes to prevent dirty water being siphoned back into the system.

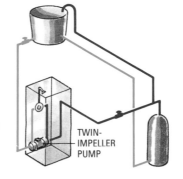

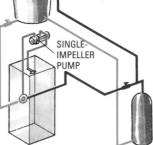

All-in-one power shower
The cold supply comes from the storage cistern, and the hot supply from the hot-water cylinder.

A separate booster pump
A typical installation with hot and cold supplies being fed through a twin-impeller pump.

High-level pump
If this is your only option, it is best to fit a single-impeller pump between the mixer and the sprayhead.

Computer-controlled showers

Computerized showers allow for the precise selection of temperature and flow rates, using a touch-sensitive control panel. Most panels also include a memory program, so that each member of a family can select their own preprogrammed ideal shower.

Far from being simply a gimmicky sales device, a computerized shower has real advantages for the disabled and for elderly people. These showers are exceptionally easy to operate – and the control panel can even be mounted outside the cubicle, so that it's possible to operate the shower on behalf of someone else.

Touch-sensitive computerized panel

☛ **SEE ALSO:** Cylinder flanges 42, Electricity 69, Electric shock treatment 80

Building a shower cubicle

Without doubt, the simplest way to acquire a shower cubicle is to install a factory-assembled cabinet, complete with tray and mixer, together with waterproof doors or a curtain to contain the spray from the sprayhead. Once you have run supply pipes and drainage, the installation is complete. However, factory-built cabinets are expensive and there is an alternative – to construct a purpose-made shower cubicle to fit the allocated space.

Choosing the site

When deciding upon the location of your shower, consider whether you can use the existing walls – or do you need new partitions to enclose the cubicle?

Freestanding
You can place the shower tray against a flat wall and either construct a stud partition on each side or surround the tray with a proprietary enclosure.

Corner site
If you position the tray in a corner of a room, then two sides of the cubicle are ready-made. Run a curtain around the tray or install a corner-entry enclosure with sliding doors. Alternatively, build a fixed side wall yourself and put either a door or a curtain across the entrance.

Built-in cupboards
To incorporate a shower cubicle un-obtrusively in a bedroom, place it in a corner, as described above, then construct a built-in wardrobe between the shower and the opposite wall.

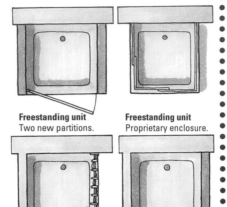

Freestanding unit
Two new partitions.

Freestanding unit
Proprietary enclosure.

Corner site
Enclosed by a curtain.

Corner site
Partition and curtain.

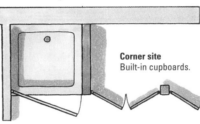

Corner site
Built-in cupboards.

Concealing the plumbing

One solution for concealing the pipes is to install a proprietary shower cubicle that has a plastic pillar in the corner, which is designed to hide the plumbing and house the mixer and adjustable sprayhead (see left).

If you erect a stud partition, then you can run the pipework between the studs. Screw exterior-grade plywood or cement-based wallboard on the inside of the frame for a tiled finish. Alternatively, use prefinished bathroom wall panelling.

Mount the shower mixer and spray-head. Finish the inside with ceramic tiles, as required, then seal the shower tray joints with mastic. You will find it easier if you connect the plumbing to the shower mixer before you enclose the outside of the partition.

If you've decided to fit decorative wall panelling, cut it to size and fix the panels, using screws and the plastic corner profiles supplied. Finally, seal all joints, including those around the edges of the tray, with waterproof mastic.

Proprietary unit
A typical kit includes a plastic corner pillar that conceals the plumbing. The kit comes complete with shower set, tray and enclosure.

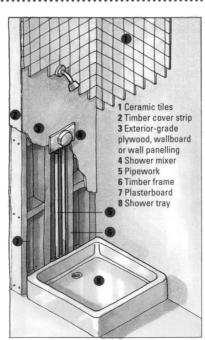

1 Ceramic tiles
2 Timber cover strip
3 Exterior-grade plywood, wallboard or wall panelling
4 Shower mixer
5 Pipework
6 Timber frame
7 Plasterboard
8 Shower tray

Running plumbing through a partition
Conceal pipework in a simple timber partition covered with ceramic tiles or panelling.

Shower trays are made from a variety of materials, but plastic trays are the most common. The relatively cheap lightweight trays tend to flex slightly in use, so it's particularly important to seal the edges carefully, using a flexible mastic (don't rely on grout). Thicker cast plastic trays are more substantial and rigid, as are ceramic trays.

The majority of shower trays are between 750 and 900mm (2ft 6in and 3ft) square. You can also buy trays that have a cut-off or rounded corner to save floor space. Larger rectangular trays provide more elbow room.

Most trays are designed to stand on the floor and have a surround that is about 150mm (6in) in height. Some have adjustable feet for levelling the tray; or even a metal underframe to raise it off the ground, providing a fall for the waste pipe. A plinth screwed across the front of the tray hides the underframe and plumbing, and provides access to the trap for servicing. Some shower trays are intended to be sunk, so that they are flush with the floor.

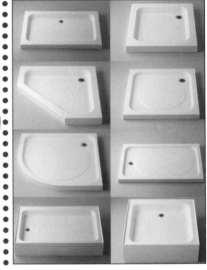

Shapes and sizes
Shower trays are between 750 and 900mm (2ft 6in and 3ft) square, usually with a surround that is about 150mm (6in) high, though this does vary. Shaped trays and ones with cut-off corners are useful where space is limited. Large rectangular trays are available to fill roomy shower cubicles.

☞ **SEE ALSO:** Enclosing a shower 41, Bathroom planning 70

Gravity-fed showers

Use the procedure below as a guide to the stage-by-stage installation of a cubicle and conventional gravity-fed shower. Ideally, you should run an independent cold supply from the storage tank; and for the hot supply, take a branch pipe directly from the vent pipe above the hot-water cylinder. Fit isolating gate valves in both supplies. Use the methods described earlier in this chapter for fitting plastic or copper supply pipes and drainage, in conjunction with the manufacturer's recommendations for the shower you are installing.

If you've decided to install an instantaneous shower in the cubicle, run both the electrical supply cable and a single 15mm (½in) pipe from the rising main through the stud partition.

Fit a non-return valve and an isolating valve in the pipe. Drill two holes in the wall just behind the shower unit for the pipe and cable. Join a threaded or compression connector to the supply pipe, whichever is appropriate for the water inlet built into the shower unit.

Read the section in this book about wiring a shower; then when you make the electrical connections, follow the manufacturer's instructions carefully.

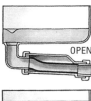

OPEN

CLOSED

Self-sealing waste valve
The flexible seal opens under waste-water pressure and then closes to form an airtight seal.

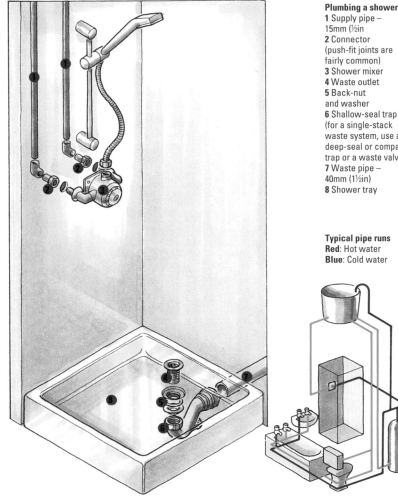

Plumbing a shower
1 Supply pipe – 15mm (½in)
2 Connector (push-fit joints are fairly common)
3 Shower mixer
4 Waste outlet
5 Back-nut and washer
6 Shallow-seal trap (for a single-stack waste system, use a deep-seal or compact trap or a waste valve)
7 Waste pipe – 40mm (1½in)
8 Shower tray

Typical pipe runs
Red: Hot water
Blue: Cold water

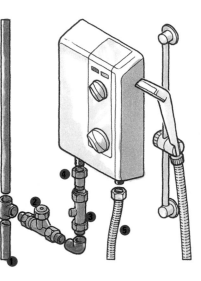

Plumbing an instantaneous shower
1 15mm (½in) pipe
2 Isolating valve
3 Non-return valve
4 Tap connector from rising main
5 Hose to sprayhead

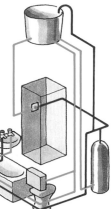

Fit the waste outlet in the shower tray and connect a shallow-seal trap, as for a bath. Alternatively, fit a compact trap that has a removable grill for easy cleaning.

Install the tray and run a 40mm (1½in) waste pipe to an outside hopper. Where the trap is to connect directly to a soil stack, rather than a hopper, you must use a conventional (deep-seal) trap or a suitable compact trap. Alternatively, you can fit a running trap or a waste valve (see far right). Check with your Building Control Officer.

To enclose a shower situated in a corner (see opposite), construct a stud partition on one side and line the inner surface with plywood or wallboard.

Cut a hole in the board for a flush-fitting shower mixer; or drill holes for the supply pipes to a surface-mounted model. Tile the inside of the cubicle with ceramic tiles, using waterproof adhesive and grout.

Fit the shower mixer and sprayhead to the tiled surface. Connect the pipework and run it back to the point of connection with the water supplies. Fit an isolating valve to each of the supply pipes, then turn off the water and make the connections.

Once the shower has been tested for leaks, cover the outside of the partition with plasterboard. Seal around the edges of the tray with a flexible silicone mastic. Finally, fit and seal the shower door.

Enclosing a shower

A shower in a cubicle or over a bath needs to be provided with some means of preventing water spraying out onto the floor. Hanging a plastic or nylon fabric curtain across the entrance is the simplest and cheapest method, but it is not really suitable for a power shower. Fit a ceiling-mounted curtain track or a tubular shower rail.

Even when a curtain is tucked into the shower tray, water always seems to escape around the sides of the curtain, or at least drips onto the floor when it is drawn aside. For a more satisfactory enclosure, use a metal-framed glass or plastic panelled unit. Hinged, sliding or concertina doors operate within an adjustable frame fixed to the top edge of the tray and the side walls. Bed the lower track onto mastic to make a waterproof joint with the tray and, once you have completed the enclosure, run a bead of mastic between the framework and the tiled walls of the shower cubicle.

Proprietary shower enclosure

☞ **SEE ALSO:** Turning off the water 6, Connecting pipes 19–27, Waste outlets 36, Shallow-seal traps 36, Running trap 37, Wiring a shower 71

Installing power showers

If you're installing a brand-new power shower, it probably pays to opt for an all-in-one model with an integral pump.

If you are merely unhappy with the performance of your existing shower, then it's much cheaper and more convenient to plumb in a separate pump.

Whichever system you choose, check that your cold-water storage capacity is typically a minimum of 115 litres (25 gallons). Some manufacturers also recommend a hot-water cylinder with a minimum 161 litres (35 gallons) capacity. Don't connect a power shower to the mains water supply.

Both types of shower need an electrical supply to drive the pump. The pump is wired to a ring main by means of a fused connection unit installed outside the bathroom. As a means of isolating the pump, use a switched fused connection unit; or, if you prefer, fit a separate ceiling-mounted double-pole switch inside the bathroom. Once connected, the shower pump switches on automatically as soon as the shower valve is operated.

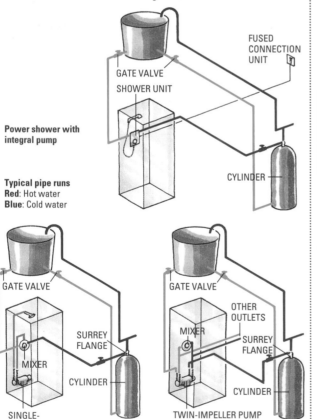

Power shower with integral pump

Typical pipe runs
Red: Hot water
Blue: Cold water

FUSED CONNECTION UNIT

GATE VALVE
SHOWER UNIT

CYLINDER

GATE VALVE

SURREY FLANGE

MIXER

CYLINDER

SINGLE-IMPELLER PUMP

GATE VALVE

OTHER OUTLETS

MIXER

SURREY FLANGE

CYLINDER

TWIN-IMPELLER PUMP

1 Single-impeller pump
Boosts ready-mixed water.

2 Twin-impeller pump
Can boost other outlets as well as a shower.

Fitting an all-in-one shower

To plumb a shower with an integral pump, you can run dedicated hot and cold supplies to the shower, as when fitting a gravity-fed shower. Alternatively, you can connect the hot-water supply directly to the cylinder by using a cylinder flange. An Essex flange is connected to the side of the cylinder (1); but to avoid cutting into the cylinder wall, fit a Surrey flange that screws into the vent-pipe connection on top of the cylinder (2). Fit gate valves in the hot and cold supplies, so you're able to isolate the shower for servicing.

The one appreciable drawback with an all-in-one shower is vibration. If you are mounting a mixer unit on a timber-frame wall, it's worth cushioning the unit on rubber tap washers slid over the fixing screws.

All tiling and grouting needs to be completed before mounting the shower on the wall.

Installing the shower

Drain the cold-water tank and drill a hole for a tank-connector fitting. Fit a gate valve close to the tank and run the pipe to the shower unit.

Turn off the cold supply to the hot-water cylinder, and then open the hot taps in the bathroom to drain a small amount of water from the cylinder. Unscrew the vent-pipe connector (3) and catch any residue of water with an old towel.

Wrap PTFE tape around the threads of the Surrey flange, then screw it into the cylinder. Connect the original vent pipe to the top of the flange and run the hot supply for the shower from the side connection (4).

Arrange the pipework at the shower end to receive connectors, making sure you have the hot and cold pipes orientated correctly for the particular unit. Open the gate valves momentarily to flush the pipes.

Following the shower manufacturer's instructions carefully, run the electrical cable to the shower, ready for connection. Unless you've had some experience of electrical wiring, have the unit wired by a qualified electrician.

Mount the shower unit, using the screws provided and taking care not to bore into pipes or cable. Connect the pipes to the unit (this is often achieved by means of simple push-fit connectors), and connect up the electrical cable to the terminal block inside the unit. Metal pipes must be bonded to earth.

Before you turn on the electricity to the pump, attach the shower hose (without the sprayhead) and use the mixer controls to run the shower fully hot then fully cold to prime both supplies. Seal around the pipes with mastic to prevent water entering the wall cavity.

Fit the cover on the unit and mount the sprayhead rail on the wall.

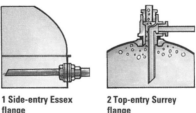

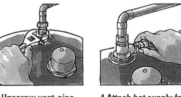

1 Side-entry Essex flange

2 Top-entry Surrey flange

3 Unscrew vent-pipe connector

4 Attach hot supply for shower

Installing a booster pump

Fitting an electric pump can improve the performance of an existing shower. If you have access to the pipe running from the mixer to the sprayhead, you can install a single-impeller pump that boosts ready-mixed hot and cold water (1). If the pipework is embedded behind tiling, install a twin-impeller pump in the supply pipes before the mixer. You can use the same twin-impeller pump to boost the supply to other outlets in the bathroom, too (2).

Positioning the pump

Place the pump somewhere convenient for servicing, perhaps on the floor under the bath, behind a screw-fixed panel – but not where it will be splashed with water. Stand it on a resilient mat or pads to reduce the noise from vibration, and do not screw it to the floor. If possible, use flexible connectors to join pipes to the pump to prevent vibration being transmitted to rigid pipework.

Connect up the pump to a switched fused connection unit (see top left). Once connected, the pump is activated automatically by flow switches.

The basic plumbing is identical to that described for installing an all-in-one shower. Flush the pipes before you switch on the pump.

☛ SEE ALSO: Turning off the water 6, Connecting pipes 20–5, 25–7, Storage tanks 49, Electricity 69, Supplementary bonding 69–70, Fused connection units 72, Electric shock treatment 80

Plumbing a bidet

Although a bidet is primarily for washing the genitals and lower parts of the body, it can double as a footbath for the elderly and for small children. Because of the stringent requirements of the Water Regulations, installing a bidet can be an expensive and time-consuming procedure. However, if you're content with the simpler version, it is just like plumbing a washbasin.

Over-rim-supply bidet

This type of bidet is simply a low-level basin. It is fitted with individual hot and cold taps or a basin mixer, and has a built-in overflow running to the waste outlet in the basin. There's one disadvantage with an over-rim bidet: the rim is cold when you sit astride it.

Rim-supply bidet

A more sophisticated bidet delivers warm water to the basin via a hollow rim. Consequently, the rim is preheated and comfortable to sit on. A special mixer set with a douche spray is fitted to this type of bidet. It incorporates the normal hot and cold valves, but a control in the centre of the mixer diverts water from the rim to the sprayhead mounted in the bottom of the basin.

Because the sprayhead is submerged when the basin is full, the Water Regulations stipulate that a rim-supply bidet must take its cold water directly from the storage tank and there must be no other connections to this cold-supply pipe. Similarly, the hot-water supply must be completely independent and connected to the vent pipe immediately above the cylinder. Check with your water supplier before installing a bidet, to make sure you comply with the regulations.

Over-rim-supply bidet *(right)*
This type of bidet is simple to install. Follow the same procedure as for a washbasin.

Rim-supply bidet *(far right)*
The installation of this type of bidet is complicated by the submerged douche spray. Independent plumbing is essential, and you will need a special mixer set to comply with the Water Regulations.

Installing a bidet

When plumbing an over-rim-supply bidet, use exactly the same procedures, pipes and connectors described for plumbing a washbasin. Fit the taps, waste outlet and trap, then use a spirit level to position the bidet before fixing it to the floor with non-corrosive screws and rubber washers. Supply the hot and cold taps with branch pipes from the existing bathroom plumbing, and take the waste pipe to the hopper or stack.

When attaching the bidet set and trap to a rim-supply appliance, follow the manufacturer's instructions. Screw the bidet to the floor before running 15mm (½in) supply pipes and a 32mm (1¼in) waste according to the Water Regulations (see left). Connect the cold supply to the tank at the same level as the existing supply pipe.

Plumbing an over-rim-supply bidet
1 Tap
2 Tap back-nut and washer
3 Tap connector
4 Supply pipe – 15mm (½in)
5 Waste outlet
6 Waste back-nut and washer
7 Trap
8 Waste pipe – 32mm (1¼in)

Space for a bidet
When planning the position of a bidet, allow sufficient knee room on each side – about 700mm (2ft 4in) overall.

Over-rim-supply bidet
Typical pipe runs.
Red: Hot water
Blue: Cold water

Rim-supply bidet
Typical pipe runs.
Red: Hot water
Blue: Cold water

Kitchen sinks

If your ambition is to re-create a period-style kitchen, you may want a reproduction Butler or Belfast fire-clay sink with a separate teak draining board. Alternatively, by way of complete contrast, you could choose a stainless-steel sink top incorporating a bowl and drainer in a single pressing. If the 'high-tech' look is not to your liking and it's colour that you're after, there are good-quality resin (plastic), enamelled and ceramic sinks available in a variety of designs and sizes.

There's a wide range of kitchen sinks, taps and accessories available for the domestic market.

Steel, enamel, resin, ceramic, double, single, plain, coloured – a bewildering choice confronts you when you are planning your kitchen. A cross section of popular sinks, accessories and taps is shown below to assist you in making your decision.

Choosing a kitchen sink

Choose the sink to make the best use of available space and to suit the style of your kitchen. If you don't have an automatic dishwasher, the kitchen sink must be large enough to cope with a considerable volume of washing-up (don't forget to allow for larger items, such as baking trays, oven racks and freezer baskets). In addition, check that the bowl is deep enough to allow you to fill a bucket from the kitchen tap.

If space allows, select a unit with two bowls. If you plan to install a waste-disposal unit, one of the bowls will need to have a waste outlet of the appropriate size (see opposite). Some sink units have a small bowl intended specifically for waste disposal.

A double drainer is another useful feature; but if there isn't enough room, allow at least some space to the side of the bowl, to avoid piling soiled and clean crockery on a single drainer.

One-piece sink tops are generally made to modular sizes to fit standard kitchen base units. However, many sinks are designed to be set into a continuous worktop – which offers greater flexibility in size, shape and, above all, positioning.

Double bowl with left-hand drainer

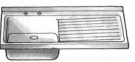

Single bowl with right-hand drainer

Inset double-bowl unit

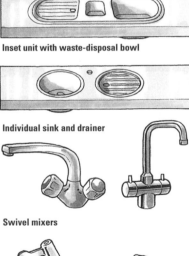

Inset unit with waste-disposal bowl

Individual sink and drainer

Kitchen taps

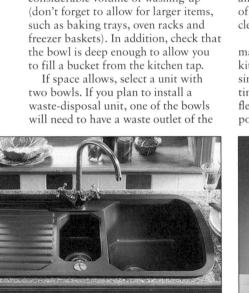

Except for being somewhat taller, kitchen taps are comparable in style to those used for washbasins. They also incorporate similar mechanisms and are fitted using the same methods.

A kitchen mixer, however, has an additional feature: drinking water is supplied to it from the rising main, whereas the hot water usually comes from the same storage cylinder that supplies all the other hot taps in the house. A sink mixer should have separate waterways to isolate the one supply from the other until the water emerges from the spout; otherwise, you must have special check valves to prevent possible contamination of your drinking water.

If you are fitting a double-bowl sink, choose a mixer with a swivelling spout. Some sink mixers have a hot-rinse spray attachment for removing food scraps from crockery and saucepans.

Continental mixer taps are supplied with small-bore malleable copper tail pipes that are screwed into the base of the taps and joined to the supply pipes by a compression-joint reducer.

Swivel mixers

Pillar tap   **Lever-operated spray**

SINK WASTE OUTLET

AIR VENT

WASTE PIPE

TRAP

Anti-siphon trap
If your trap gurgles as the sink empties, you could replace it with an anti-siphon trap. This type of trap draws in air to break the vacuum in the waste pipe.

Accessories for a kitchen sink

You can buy a variety of accessories to fit most kitchen sinks, including a hardwood or laminated-plastic chopping board that drops neatly into the rim of the bowl or drainer, and a selection of plastic-dipped wire baskets for rinsing vegetables or draining crockery.

Pump-action dispensers for soap and washing-up liquid rid the sink of plastic bottles and soap dishes.

Chopping boards

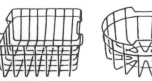

Wire baskets

☛ **SEE ALSO: Installing a sink 45**

Installing a sink

WASTE-DISPOSAL UNITS

Installing a kitchen sink is much the same as fitting a washbasin or vanity unit. All except ceramic sinks will require a combined overflow/waste outlet, like a bath. It pays to fit a tubular trap to a sink, because a bottle trap blocks too easily.

A waste-disposal unit provides a hygienic method of dealing with soft food scraps – reserving the kitchen wastebin for dry refuse and bones.

The unit houses an electric motor that drives steel cutters, which grind up the food scraps into a fine slurry to be washed into the yard gully or soil stack. A continuous-feed model is operated by a manual switch: scraps are then fed into it while the cold tap is running. To prevent the unit being switched on accidentally, a batch-feed model cannot be operated until a removable plug is inserted in the sink waste outlet.

Waste-disposal units are generally designed to fit an 89mm (3½in) outlet in the base of the sink bowl. A special cutter can be hired to adapt a standard stainless-steel or plastic sink.

With a sink waste outlet and seal in position, clamp a retaining collar to the outlet from under the sink. Bolt or clip the unit housing to the collar: every unit is supplied with individual instructions.

The waste outlet from the unit itself fits a standard sink trap (not a bottle trap) and waste pipe. If the waste pipe runs to a yard gully, make sure it passes through the covering grid (see left).

Wire the unit to a switched fused connection unit mounted above the worktop, positioning it so that it is out of the reach of children. Identify the switch to avoid accidental operation.

Cutting a hole for a waste-disposal unit
The supplier of the waste-disposal unit (or possibly a tool-hire company) will rent you a special cutter to convert an existing sink. The cutter can't be used on a ceramic or enamel sink.

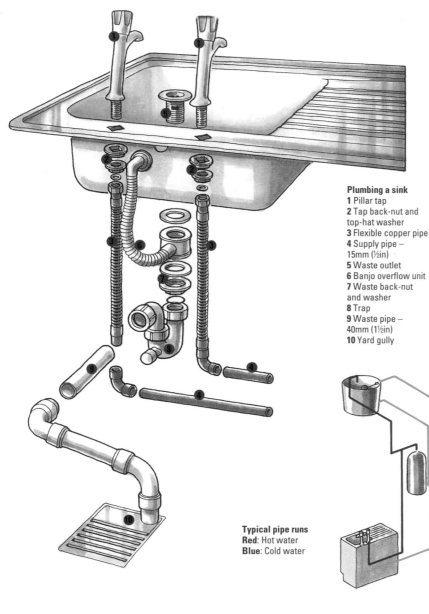

Plumbing a sink
1 Pillar tap
2 Tap back-nut and top-hat washer
3 Flexible copper pipe
4 Supply pipe – 15mm (½in)
5 Waste outlet
6 Banjo overflow unit
7 Waste back-nut and washer
8 Trap
9 Waste pipe – 40mm (1½in)
10 Yard gully

Typical pipe runs
Red: Hot water
Blue: Cold water

Waste-disposal unit
Units differ in detail, but the illustration shows the components typically used to clamp a waste-disposal unit to a sink.
1 Sink waste outlet
2 Gasket
3 Back-up ring
4 Collar
5 Snap ring
6 Unit housing
7 Cutters
8 Waste outlet
9 Trap

A fast and hygienic way to dispose of soft food scraps

Plumbing the sink

Fit the taps and the overflow/waste outlet to the new sink before you place the sink in position.

Turn off the water supply to the taps, then remove the old sink by dismantling the plumbing. Remove the old pipework unless you plan to adapt it.

Clamp the new sink to its base unit or worktop, using the fittings provided; then, if needed, seal the rim of the sink. Run a 15mm (½in) cold-water supply pipe from the rising main, and a branch pipe of the same size from the nearest hot-water pipe. Fit miniature isolating valves in both of the supply pipes and connect them to the taps with flexible copper-tap connectors.

Fit the trap and run a 40mm (1½in) waste pipe through the wall behind the base unit to the yard gully. According to current Water Regulations, the pipe has to pass through the grid covering the gully but must stop short of the water in the gully trap. You can adapt an existing grid quite easily by cutting out one corner with a sharp hacksaw.

☞ **SEE ALSO:** Wiring Regulations 6, 81, Connecting pipes 19–27, Tap connectors 24, Washbasins 31, Fused connection units 72, Overflow pipe 81

45

Dishwashers and washing machines

Nowadays dishwashers and washing machines are to be found in most kitchens or utility rooms. Made to standard sizes to conform with kitchen fitments, they fit neatly under a work surface and are attached by flexible hoses to a dedicated waste pipe or to the waste from the kitchen or utility-room sink.

Automatic machines should have permanent supply and waste systems. Dishwashers need a cold supply only, whereas washing machines may be hot-and-cold fill. Washing machines that are supplied with hot water provide a faster washing cycle; and they may be more economical to run, depending on how you heat your water. Any retailer will be happy to advise you.

Appliance valves
Typical valves used to connect dishwashers and washing machines to the water supply.

In-line valve

Right-angle valve

T-piece valve

Water pressure

The instructions accompanying the machine should indicate what water pressure is required. If the machine is installed upstairs, make sure the drop from the storage tank to the machine is big enough to provide the required pressure. In a downstairs kitchen or utility room there is rarely any problem with pressure, especially if you can take the cold water from the mains supply at the sink. However, check with your water supplier if you want to connect more than one machine.

Plumbing a washing machine
1 Supply pipe – 15mm (½in)
2 Appliance valve
3 PVC inlet hoses
4 Machine inlets
5 Outlet hose
6 Standpipe
7 Trap
8 Waste pipe – 40mm (1½in) – to gully

Running the supply

Washing machines and dishwashers are supplied with PVC hoses to link the water inlets at the back of the appliance to special miniature valves connected to the household plumbing. Using these valves, you can turn off the water when you need to service a machine, without having to disrupt the supply to the rest of the house. There are a number of valves to choose from. Select the type that provides the most practical method of connecting to the plumbing, depending on the location of the machine in relation to existing pipework.

Self-bore valves

When 15mm (½in) cold and hot pipes run conveniently behind or alongside the machine, use a valve that will bore a hole in the pipe without your having to turn off the water and drain the system. Each valve is colour-coded for hot or cold, and has a threaded outlet for the standard machine hose. Self-bore valves are not approved by all water suppliers because the small disc of metal they cut from the pipe may restrict the flow of water. In practice, this hardly ever happens.

To fit a valve, screw the backplate to the wall behind the pipe. Place the saddle with its rubber seal over the pipe. Before screwing the saddle to the backplate **(1)**, ensure that the seal in the saddle is positioned correctly.

Make sure the valve is turned off, then screw it into the saddle **(2)**. As you insert the valve, the integral cutter bores a hole in the pipe. With the valve in the vertical position, tighten the adjusting nut with a spanner **(3)**; then connect the hose to the valve outlet **(4)**.

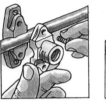 

1 Fit the saddle 2 Insert the valve

3 Tighten the nut 4 Attach the hose

Running branch pipes

If you have to extend the plumbing to reach the machine, take branch pipes from the hot and cold pipes supplying the kitchen taps. Terminate the branch pipes at a convenient position close to the machine, and fit a small appliance valve (see far left) that has a standard compression joint for connecting to the pipework and a threaded outlet for the machine hose. Before fitting this type of valve, turn off the water and drain the system in the normal way. When you have restored the supply, open the valve by turning the control level to align with the outlet.

☞ **SEE ALSO:** Draining the system 8, Connecting pipes 19–27, Storage tanks 49

Preventing a flood

The outlet hose from a dishwasher or washing machine must be connected to a waste system that will discharge the dirty water into either a yard gully or a single waste stack – not into a surface-water drain, where detergents could pollute rivers.

Standpipe and trap

The standard method, approved by all water suppliers, employs a vertical 40mm (1½in) plastic standpipe attached to a deep-seal trap (see opposite).

Most plumbing suppliers stock the standpipe, trap and wall fixings as a kit. The machine hose fits loosely into the open-ended pipe, so that dirty water won't be siphoned back into the machine. The machine manufacturer's instructions should tell you how to position the standpipe; in the absence of advice, ensure that the open end is at least 600mm (2ft) above the floor.

Cut a hole through the wall and run the waste pipe to a gully; or use a pipe boss to connect the waste to a drainage stack. Allow a minimum fall of 6mm (¼in) for every 300mm (1ft) of pipe run.

Draining to a sink trap

You can drain a washing machine to a sink trap that has a built-in spigot (1), but you should insert an in-line anti-siphon return valve in the machine's outlet hose. This is a small plastic device with a hose connector at each end (2). In order to drain a washing machine and dishwasher together, you will need a dual-spigot trap.

1 Sink trap with drainage spigot

2 In-line anti-siphon hose valve

Overflowing dishwashers and washing machines can cause a great deal of damage in just a few minutes – particularly if the appliance is plumbed into an upstairs flat and the water is able to find its way through a multi-storey building.

Air-inlet valves

Most overflows occur simply because the water backs up the waste pipe and spills out over the standpipe or sink.

A sealed waste system succeeds in overcoming this problem – since it does away with the air gap that allows the water to overflow. The anti-vacuum function is formed, instead, by a fitting that incorporates a small air-inlet valve, which stops the waste pipe siphoning the machine. The discharge hose from the machine is connected to the nozzle of the vent fitting, and a length of 40mm (1½in) waste pipe is inserted between the fitting and the washing machine trap under the sink.

Anti-siphon devices

The standpipe-and-trap method of draining domestic appliances prevents back-siphonage by venting the pipe to the air, but there are other ways to deal with the problem. If an existing 32 or 40mm (1¼ or 1½in) waste pipe runs behind the machine, for example, you can attach a hose connector that incorporates a non-return valve to eliminate reverse flow. Connectors are available with short spigots (1), or can be attached to a standpipe.

Connecting to the waste pipe
Clamp the saddle over the waste pipe (2), then use the cutter supplied with the fitting to bore a hole in the pipe, with the saddle acting as a guide (3).

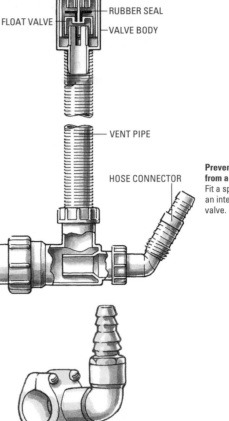

AIR HOLE
RUBBER SEAL
FLOAT VALVE
VALVE BODY
VENT PIPE
HOSE CONNECTOR

Preventing an overflow from a standpipe
Fit a special vent with an integral air-inlet valve.

1 Short-spigot anti-siphon connector
This type of connector is clamped to a waste pipe that runs behind the machine.

2 Clamp the saddle over the waste pipe

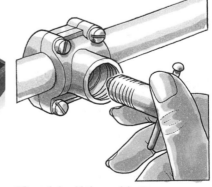

3 Bore a hole with the special cutter

☞ **SEE ALSO: Connecting pipes 19–27**

Water softeners

Harmful impurities are removed from water before it is supplied to our homes, but minerals absorbed from the ground are still present and it's the concentration of these that determines whether our water is hard or soft. Rocky terrain gives rise to surface-run water, which is naturally soft – whereas in areas of the country where water runs through the ground, rather than over it, the higher mineral content produces hard water.

Water softener
A domestic unit, which fits neatly beneath the worktop, requires topping up with salt.

Hard-water scale

Mineral salts are deposited in the form of hard scale on the inside of pipes, tanks and, especially, hot-water cylinders. If the concentration of minerals is very high, scale will eventually block pipework and can insulate heating elements to such an extent that their efficiency is reduced by anything from 15 to 70 per cent.

The more obvious consequences of hard water are the discoloration of baths and basins, blocked sprayheads, blemished stainless-steel surfaces and furred-up kettles. Most people resign themselves to living with these effects – but they can be reduced, or even eliminated altogether, by installing a water softener.

Domestic water softeners

Water softeners work on the principle of ion exchange. The incoming water flows through a compartment containing a synthetic resin that absorbs scale-forming calcium and magnesium ions and releases sodium ions in their place.

After a period of about three or four days, the resin is unable to absorb any more mineral salts and the softener automatically flushes the compartment with a saline solution to regenerate the resin. Topping up with salt is required at intervals of perhaps two to three months. The softener is fitted with a timer so that you can program regeneration when water consumption is at its lowest, usually during the early hours of the morning.

The unit must be connected to the rising main at the point where the water supply enters the house. For this reason, domestic softeners are usually designed to fit under a kitchen worktop.

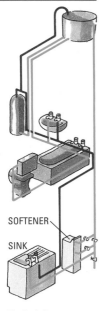

Typical pipe runs
A domestic system incorporating a softener.
Red: Hot water
Blue: Cold water

SOFTENER
SINK

Installing a water softener

Installing a water softener may appear to be fairly complicated since it involves a great deal of joint making – both to fit the valves and branch pipes that supply and bypass the softener and to include the fittings that are necessary to comply with the Water Regulations.

The bypass assembly allows for the unit to be isolated for servicing while maintaining the supply of water to the rest of the house. In addition, you must install a branch pipe before the assembly, in order to supply unsoftened drinking water to the kitchen sink. Supply your garden tap (see top right) from the same pipe – there's no need to waste softened water on the garden.

Install a non-return valve in the system, to prevent the reverse flow of salty water. A pressure-reducing valve may also be required (check with your water supplier). You will need a drain-cock, in order to empty the rising main. Some manufacturers supply an install-ation kit that includes all the necessary equipment. You will have to provide drainage in the form of a standpipe and trap, as for a washing machine.

Wire the water softener to a switched fused connection unit that contains a 3amp fuse.

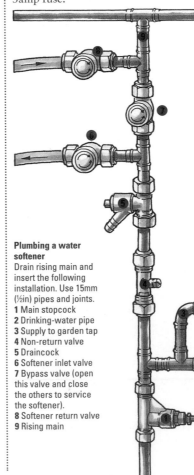

Plumbing a water softener
Drain rising main and insert the following installation. Use 15mm (½in) pipes and joints.
1 Main stopcock
2 Drinking-water pipe
3 Supply to garden tap
4 Non-return valve
5 Draincock
6 Softener inlet valve
7 Bypass valve (open this valve and close the others to service the softener).
8 Softener return valve
9 Rising main

A bib tap situated on an outside wall is convenient for attaching a hose for a lawn sprinkler or for washing the car. To comply with the Water Regulations, a double-seal non-return (check) valve must be incorporated in the plumbing, to prevent contaminated water being drawn back into the system. Provide a means of shutting off the water and draining the pipework during winter, and keep the outside pipe run as short as possible.

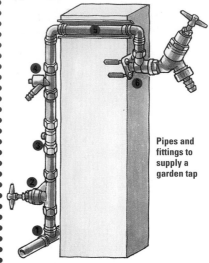

Pipes and fittings to supply a garden tap

Turn off and drain the mains supply. Fit a T-joint (**1**) to run the supply to the tap. Run a short length of pipe to a convenient position for another stopcock (**2**) or miniature valve, and for the non-return valve (**3**) if the tap doesn't include one, making sure that the arrows marked on both fittings point in the direction of flow. Fit a draincock (**4**) after this point. Run a pipe through the wall inside a length of plastic overflow (**5**), so that any leaks will be detected quickly and will not soak the masonry. Wrap PTFE tape around the bib-tap thread, then screw it into a wall plate attached to the masonry outside (**6**).

A suitably robust bib tap for use outdoors

☞ **SEE ALSO:** Draining the system 8, Connecting pipes 19–27, Washing machines 46–7, Fused connection units 72, PTFE tape 81

Storage tanks

The cold-water storage tank, or cistern, normally situated in the roof space, supplies the hot-water cylinder and all the cold taps in the house, other than the one in the kitchen that is used for drinking water. An old house may still have a galvanized-steel tank that has been in service since the house was built. But eventually this will corrode and, although it's possible to patch it up temporarily, it makes sense to replace it before a serious leak develops. A circular 227 litre (50 gallon) polythene tank is a popular replacement, because it can be folded to pass through a narrow hatch to the loft.

Bylaw 30 kits

Make sure your new tank is supplied with a Bylaw 30 kit, to keep the water clean. This is a requirement of all water suppliers. The kit includes a close-fitting lid that excludes light and insects, and is fitted with a screened breather and a sleeved inlet for the vent pipe. In addition, there should be an overflow-pipe assembly that is screened to prevent insects crawling into the tank, a reinforcing plate to stiffen the cistern wall around the float valve, and an insulating jacket.

Removing an old tank

Switch off all water-heating appliances, then close the stopcock on the rising main. Drain the storage tank by opening the cold taps in the bathroom.

Bail out the remaining water in the bottom of the tank, then use a spanner to dismantle the fittings connecting the float valve, distribution pipes and overflow to the tank. Use a little penetrating oil if the fittings are stiff with corrosion.

The tank may have been built into the house before the roof was completed, in which case it's unlikely to pass through the hatch. Just pull it to one side. If you need the space, it is possible to cut the tank up, using an angle grinder. Wear a mask, gloves, goggles and ear defenders while you work.

Prepare a firm base for the new tank by nailing stout planks across the joists, or lay a platform made from plywood 18mm (¾in) thick.

Plumbing a new tank

Once the new tank is in place, you can set about connecting the numerous pipes and fittings that are required.

Fitting the float valve
A float valve shuts off the flow of water from the rising main when the tank is full. Cut a hole for the float valve 75mm (3in) below the top of the tank. Slip a plastic washer onto the tail of the valve and pass it through the hole. Slide the reinforcing plate onto the tail, followed by another washer and a fixing nut, then tighten the fitting with the aid of two spanners.

Screw a tap connector onto the float valve, ready for connecting to the 15mm (½in) rising main.

Connecting the distribution pipes
The 22mm (¾in) pipes running to the cylinder and cold taps are attached by means of tank connectors – threaded inlets with a compression fitting for the pipework. Drill a hole for each tank connector, about 50mm (2in) above the bottom of the tank. Push the fittings through each hole, with one polythene washer on the inside. Wrap a couple of turns of PTFE tape around the threads, then fit the other washer. Screw the nut on, holding the tank connector to stop it turning. Don't overtighten the nut – or you will damage the washer, causing it to leak.

Take the opportunity to fit a gate valve to each distribution pipe, so you can cut off the supply of water without having to empty the tank.

Connecting the overflow
Drill a hole 25mm (1in) below the level of the float-valve inlet for the threaded connector of the overflow-pipe assembly. Pass the connector through the hole, fit a washer, and tighten its fixing nut on the inside of the tank. Fit the dip pipe and insect filter.

Attach a 22mm (¾in) plastic overflow pipe to the assembly. Run the pipe to the floor, then to the outside of the house, maintaining a continuous fall. The pipe must emerge in a conspicuous position, so that an overflow can be detected immediately. Clip the pipe to the roof timbers.

Modifying existing plumbing
Modify the rising main and distribution pipes to align with their fittings, then connect them with compression fittings. (Don't use soldered joints near a plastic tank.) Clip all the pipework securely to the joists.

Open the main stopcock and check for leaks as the tank fills. Adjust the float arm to maintain a water level 25mm (1in) below the overflow outlet.

Adapt the vent pipe from the hot-water cylinder to pass through the hole in the lid. Finally, insulate the tank and pipework – but make sure there is no loft insulation under the cistern, as this will prevent warmth rising from below.

Tank cutters
Hire a tank cutter to bore holes in the tank for pipework. Some cutters are adjustable, so you can drill holes of different diameters. An alternative is to use a hole saw clamped to a drill bit.

Hole saw

Adjustable cutter

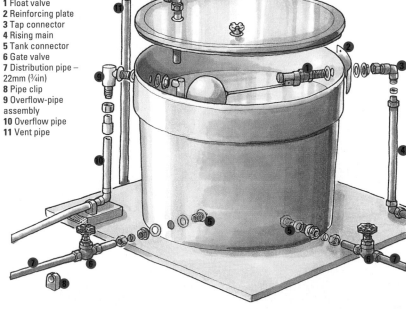

Plumbing a tank.
1 Float valve
2 Reinforcing plate
3 Tap connector
4 Rising main
5 Tank connector
6 Gate valve
7 Distribution pipe – 22mm (¾in)
8 Pipe clip
9 Overflow-pipe assembly
10 Overflow pipe
11 Vent pipe

☞ **SEE ALSO:** Gate valve 8, Adjusting a float arm 14, Compression joints 20, Tap connectors 24, Hot-water cylinders 50–1, Float valves 81

Vented hot-water cylinders :HOT-WATER CYLINDERS

In most houses, the hot water is heated and stored in a large copper cylinder situated in the airing cupboard. Cold water is fed to the base of the cylinder from the cold-water storage tank housed in the loft. As the water is heated, it rises to the top of the cylinder, where it is drawn off via a branch from the vent pipe to the hot taps. When the hot water is run off, it is replaced by cold water at the base of the cylinder, ready for heating.

The vent pipe itself runs back to the loft, where it passes through the lid of the cold-water storage tank, with its open end just above the level of the water. The vent pipe provides a safe escape route for air bubbles and steam, should the system overheat.

When water is heated, it expands. The vent pipe accommodates some of this expansion, but much of the excess water is forced back up the cold-feed pipe into the cold-water storage tank.

The capacity of domestic cylinders normally ranges from about 114 litres (25 gallons) to 227 litres (50 gallons), although it is possible to obtain bigger cylinders to meet the requirements of a large family. A cylinder with a capacity of between 182 and 227 litres (40 and 50 gallons) will store enough hot water to satisfy the needs of an average family for a whole day.

Some cylinders are made from thin, uninsulated copper and need to have a thick lagging jacket to reduce heat loss. However, for better performance use a Kite-marked factory-insulated cylinder that is precovered with a thick layer of foamed polyurethane. Although more expensive, they are a good investment.

Typical pipe runs
Red: Hot water
Blue: Cold water

Direct water heating by means of a boiler

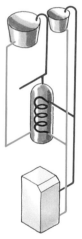

Indirect water heating employs the central-heating boiler

Methods of heating water

There are two different methods of heating the water in a vented hot-water cylinder: either directly – usually by means of electric immersion heaters – or indirectly by a heat exchanger connected to the central-heating system.

Direct heating
Water heating can be accomplished solely by means of electric immersion heaters – either a single-element or double-element heater is fitted in the top of the cylinder or there may be two individual side-entry heaters.

An alternative is for the water to be heated in a boiler, the sole purpose of which is to provide hot water for the cylinder. A cold-water pipe runs from the base of the cylinder to the boiler, where the water is heated; and it then returns to the top half of the cylinder.

Both methods are known as direct systems. In practice, a boiler-heated cylinder is generally fitted with an immersion heater as well, so that hot water can be supplied independently during the summer, when using the boiler would make the room where it is situated uncomfortably warm.

Indirect heating
When a house is centrally heated with radiators fed by a boiler, the water in the cylinder is usually heated indirectly by a heat exchanger.

Hot water from the boiler passes through the exchanger (a coiled tube within the cylinder), where the heat is transmitted to the stored water. The heat exchanger is part of a completely self-contained system, which has its own feed-and-expansion tank (a small storage tank in the loft) to top up the system. An open-ended vent pipe terminates over the same small tank.

The whole system is known as the primary circuit, and the pipes running from and back to the boiler are known as the primary flow and return. An indirect system is often supplemented with an immersion heater, to provide hot water during the summer months.

Changing a cylinder

You may wish to replace an existing cylinder because it has sprung a leak, or because a larger one will allow you to take full advantage of cheap night-time electricity by storing more hot water. A simple replacement can some-times be achieved without modifying the plumbing, but you'll have to adapt the pipework to fit a larger cylinder.

If you plan to install central heating at some point in the future, you can plumb in an indirect cylinder fitted with a double-element immersion heater and simply leave the heat-exchanging coil unconnected for the time being.

First switch off and disconnect any immersion heaters from the electrical supply, then drain the cylinder and pipe-work. Using a special spanner (available from a tool-hire outlet), unscrew the immersion heaters. Disconnect all the pipework, springing it out of the way while you remove the cylinder.

Place the new cylinder in position and check the existing pipework for alignment. Modify the pipes as need be, then make the connections, using PTFE tape to ensure that the threaded joints are watertight. Fit a draincock to the feed pipe from the tank, if there isn't one already installed.

With the fibre sealing washer in place, wrap PTFE tape around the thread of the immersion heater and screw it into the cylinder. Connect the immersion heater to the electrical supply, then fill the system and check for leaks before you attempt to heat the water. Check for leaks again when the water is up to temperature.

Direct cylinder
1 Vent pipe
2 Hot-water branch pipe
3 Lower immersion heater (provides hot water using cheaper night-rate electricity)
4 Upper immersion heater (used for daytime top-up heating only)
5 Cold-feed pipe
6 Draincock

Indirect cylinder
1 Vent pipe
2 Back-up immersion heater
3 Flow from boiler
4 Heat exchanger
5 Return to boiler
6 Draincock
7 Cold feed from tank

☛ **SEE ALSO:** Draining the system 8, Connecting pipes 19–27, Compression joints 20, Wiring side-entry heaters 73, Wiring immersion heaters 73, Electric shock treatment 80, PTFE tape 81

Unvented cylinders

A thermal-store cylinder reverses the indirect principle. Water heated by a central-heating boiler passes through the cylinder and transfers heat, via a highly efficient coiled heat exchanger, to mains-fed water supplying hot taps and showers. An integral feed-and-expansion tank is normally built on top of the cylinder.

When the system is working at maximum capacity, the mains-fed water is delivered at such a high temperature that cold water must be added via a thermostatic mixing valve plumbed into the outlet supplying taps and showers. As the cylinder is exhausted, less cold water is added. The thermal-store system provides mains-pressure hot water throughout the house, dispenses with the need for a cold-water storage tank in the loft, and increases the efficiency of the boiler.

A valve is needed to prevent the heat from the cylinder 'thermo-siphoning' (gravity circulating) around the central-heating system. This can be a motorized valve or a simple mechanical gravity-check (non-return) valve that is opened by the force of the central-heating pump.

As with all open-vented systems, the feed-and-expansion tank determines the head of water, and radiators must be lower than the tank in order to be filled with water. When the tank is combined with the cylinder, it needs to be situated on the top floor of the house in order to provide central heating throughout the building. If that is impossible, install a tankless thermal-store cylinder and fit a conventional feed-and-expansion tank in the loft.

An unvented cylinder supplies mains-pressure hot water throughout the house. This is achieved by connecting the cylinder directly to the rising main. Most manufacturers recommend a 22mm (¾in) incoming pipe, but in practice a 15mm (½in) main at high pressure is normally adequate. An unvented cylinder can be heated directly, using immersion heaters, or indirectly, provided you are not using a solid-fuel boiler.

There are no storage tanks, feed-and-expansion tanks or open-vent pipes associated with unvented cylinders. Instead, a diaphragm inside a pressure vessel mounted on top of the cylinder flexes to accommodate expanding water. If the vessel fails, an expansion-relief valve protects the system by releasing water via a discharge pipe.

There are several other safety devices associated with unvented cylinders. A normal thermostat should keep the temperature of the water in the cylinder below 65°C (150°F). If it reaches 90°C (195°F), then a second thermostat will either switch off the immersion heaters or shut off the water supply from the boiler. Finally, if it should get as hot as 95°C (205°F), a temperature-relief valve opens and discharges water outside.

Bylaws and regulations

The installation of an unvented hot-water cylinder needs to comply with both the Water Regulations and the Building Regulations. It has to include all the necessary safety devices and be installed by a competent fitter, such as those registered with the Institute of Plumbing, the Construction Industry Training Board, or the Association of Installers of Unvented Hot Water Systems (Scotland and Northern Ireland). Have the installation serviced regularly by a similarly qualified fitter, to make sure all the equipment remains in good working order.

You must notify the water company and your local Building Control Office of your intention to install an unvented hot-water cylinder.

Thermal-store cylinder
1 Integral feed-and-expansion tank
2 Heat-exchanger
3 Supply pipe to hot taps/shower
4 Thermostatic mixing valve
5 Expansion vessel
6 Mains feed
7 Space-heating flow
8 Space-heating return
9 Boiler flow
10 Boiler return

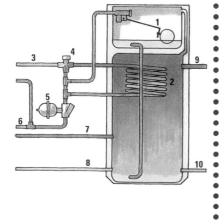

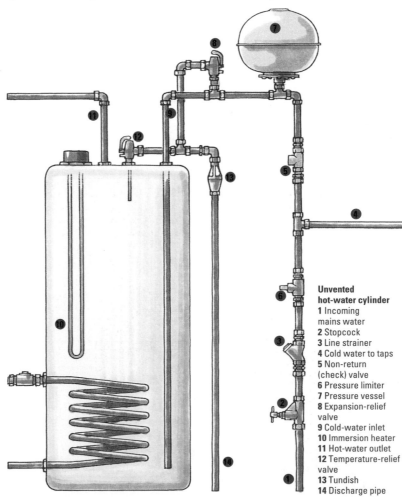

Unvented hot-water cylinder
1 Incoming mains water
2 Stopcock
3 Line strainer
4 Cold water to taps
5 Non-return (check) valve
6 Pressure limiter
7 Pressure vessel
8 Expansion-relief valve
9 Cold-water inlet
10 Immersion heater
11 Hot-water outlet
12 Temperature-relief valve
13 Tundish
14 Discharge pipe

☞ **SEE ALSO: Storage tanks 49, Wet central heating 53**

Solar heating

Saving energy is a priority for all of us if we are to prevent further damage to our environment from the effects of carbon dioxide.

Point-of-use water heaters help in a small way, as they consume energy for short periods only. However, the systems that have been developed to harness solar energy offer a more effective alternative for heating domestic water. In contrast to the demand for space heating, which varies according to the season, hot water is required constantly throughout the year – and is therefore well suited to heating with solar energy.

Using solar energy to heat water

The idea of using the sun to provide free, non-polluting energy for heating water has always appealed to energy-conservationists but has yet to become widely accepted. However, with the development of the new generation of evacuated-heat-pipe solar collectors, it is now possible to heat domestic hot water effectively and economically.

From the late spring through to early autumn, this type of system can produce sufficient hot water for the average house – even when the sky is overcast. During the winter, the solar collectors provide useful 'preheat' that reduces the time it takes a boiler to heat water, thereby saving energy.

There are a number of companies that supply solar collectors for heating water, plus all the controls and pipe-work required to complete the job. If you carry out the plumbing yourself, the payback on the investment will be that much greater.

Mount collectors on a south-facing roof

A basic system
Most systems for supplying domestic hot water will require solar collectors that cover about 4sq m (4sq yd) of roof space. In order to trap maximum heat from the sun, the collectors should be mounted on a pitched roof and face in a southerly direction. Solar collectors can be fitted, with minimal structural alterations, to almost any building; and planning approval is rarely required.

The most common way of utilizing solar energy to boost an existing water-heating system is to feed the hot water from the collectors to a second heat exchanger fitted inside your hot-water cylinder. This usually means replacing the cylinder with a dual-coil model.

An alternative technique is to plumb in a second well-insulated cylinder, which will 'preheat' the water before it is passed on to the main storage cylinder. This may involve raising the cold-water storage tank in order to feed the new preheat cylinder.

Controls
A pump is needed to circulate the water from the collectors to the cylinder coil and back to the collectors. A pro-grammable thermostat, which operates the pump, senses when the panels are hotter than the water in the cylinder.

Small instantaneous water heaters are used to provide hot water at the point where it is required, usually beside a sink or basin. A 3kW model, suitable for mounting above a sink, is wired to a fused connection unit containing a 13amp fuse. The unit must be out of reach of water splashes from the sink, so if necessary fit a flex outlet near the heater and run a cable from there to the connection unit.

A 7kW heater needs a 45amp radial circuit, similar to the one for a shower, though in a kitchen you can use a wall-mounted double-pole switch to connect it, instead of a ceiling-mounted switch.

Electric point-of-use water heaters are often designed to fit inside a cup-board or vanity unit beneath a sink or basin. You can install one of these heaters yourself, provided that it has a capacity of less than 9 litres (16 pints). Follow the manufacturer's instructions precisely, and fit a pressure-limiting valve and a filter (both of these are supplied as a kit). Also, make sure that the safety vent pipe discharges hot water to a place outside where it won't endanger anyone.

Electric water heaters are supplied directly from the mains by means of a 15mm (½in) pipe.

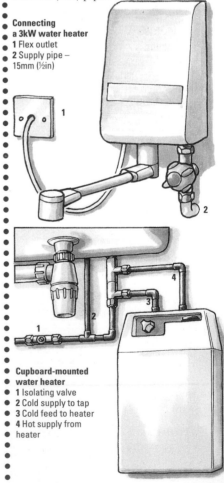

Connecting
a 3kW water heater
1 Flex outlet
2 Supply pipe –
15mm (½in)

Cupboard-mounted
water heater
1 Isolating valve
2 Cold supply to tap
3 Cold feed to heater
4 Hot supply from
heater

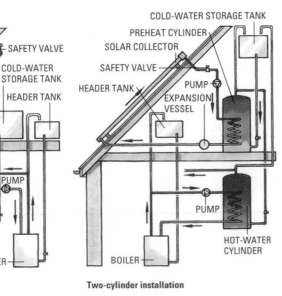

Dual-coil installation

Two-cylinder installation

☞ SEE ALSO: Wiring Regulations 6, 39, 69, 81, Connecting pipes 19–27, Hot-water cylinders 50–1

Wet central heating

Open-vented systems

The most popular form of wet central heating is the two-pipe open-vented system – in which water is heated by a boiler and pumped through small-bore pipes to radiators or convector heaters, where the heat from the water is released into the rooms. The water then circulates back to the boiler for reheating, using natural gas, bottled gas (propane), oil, electricity, or a solid fuel such as anthracite.

The control of such systems can be extremely flexible. Thermostats and valves allow the output of the individual heat emitters to be adjusted automatically, and parts of the system can be shut down when rooms are not used.

This type of system can be used to heat the domestic hot-water supply, as well as the house itself. Some older systems employ gravity circulation to heat the hot-water storage cylinder but incorporate a mechanical pump to force the water around the radiators. In most modern systems, a similar pump propels the water to the cylinder and radiators via diverter valves.

Sealed systems

A sealed system is an alternative to the traditional open-vented method. Water is fed into the system via a filling loop, which is temporarily connected to the mains. The loop incorporates a non-return valve to prevent contamination of mains drinking water. In place of a feed-and-expansion tank (see top right), a pressure vessel containing a flexible diaphragm accommodates the expansion of the water as the temperature rises. Should the system become over-pressurized, a safety valve discharges some of the water.

A sealed central-heating system offers certain advantages over an open-vented system. There is less likelihood of corrosion and, since the system runs at a relatively high temperature, the radiators can be smaller. Also, because the system is supplied with water under mains pressure, there is no necessity for radiators to be below a feed-and-expansion tank installed in the loft – so radiators can be placed anywhere in the house, including in the loft itself.

On the negative side, sealed systems must be completely watertight – since there is no automatic top up – and they have to be made with costly high-quality components to prevent pressure loss. A boiler with a high-temperature cutout is required, in case the ordinary thermostat fails. Also, radiators get very hot.

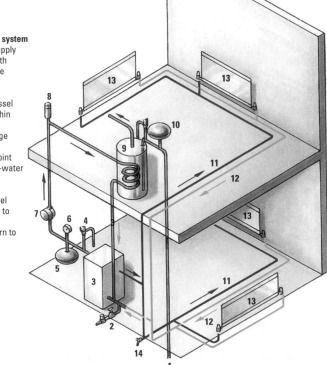

Sealed heating system
1 Cold mains supply
2 Filling loop with non-return valve
3 Boiler
4 Safety valve
5 Expansion vessel (sometimes within boiler)
6 Pressure gauge
7 Pump
8 Air-release point
9 Unvented hot-water cylinder
10 Hot-water expansion vessel
11 Heating flow to radiators
12 Heating return to boiler
13 Radiators
14 Draincock

Open-vented system
The water heated by the boiler (1) is driven by a pump (2) through a two-pipe system to the radiators (3) or special convector heaters, which give off heat as the hot water flows through them, gradually warming the rooms to the required temperature; the water then returns to the boiler to be reheated. A cistern known as a feed-and-expansion tank (4), situated in the loft, keeps the system topped up and takes the excess of water created by the system overheating. The hot-water cylinder (5) is heated by gravity circulation. In the diagram, red indicates the flow of water from the pump and blue shows the return flow.

● **One-pipe systems**
In an outdated one-pipe system, heated water is pumped around the perimeter of the house through a single large-bore pipe that forms a loop. Flow and return pipes divert hot water to each radiator by means of gravity circulation. Larger radiators may be required at the end of the loop in order to compensate for heat loss. A one-pipe system incorporates a feed-and-expansion tank and a hot-water circuit similar to those used for conventional two-pipe systems.

Central-heating boilers

Technological improvements have made it possible to produce central-heating boilers much smaller than their predecessors, though no less efficient. Today, gas and oil are still the most popular fuels because, despite advances in solid-fuel techno-logy, the dirt and inconvenience associated with solid fuels can't be ignored or overcome. Wood-burning boilers were popular for a while – but, realistically, wood is best suited to room-heating stoves, perhaps with a small back boiler to provide hot water, rather than as a fuel for central heating.

● Gas installers
Gas boilers must be installed by competent fitters registered with CORGI (Council for Registered Gas Installers). Check, also, that your installer has the relevant public-liability insurance for working with gas.

● Boiler flues
All boilers need some means of expelling the combustion gases that result from burning fuel. Frequently this is effected by connecting the boiler to a conventional flue or chimney that takes the gases directly to the outside.

Alternatively, some boilers, known as room-sealed balanced-flue boilers, are mounted on an external wall and the flue gases are passed to the outside through a short horizontal duct. Balanced-flue ducts are divided into two passages – one for the outgoing flue gases, and the other for the incoming air needed for efficient combustion.

All boilers can be connected to a conven-tional flue, but gas and oil-fired boilers are also made for balanced-flue systems. If the boiler is fan-assisted, it can be mounted at a distance of up to 3m (9ft 9in) from the balanced-flue outlet.

Heating requirements

The capacity (heat output) of the boiler needed to satisfy your requirements can be calculated by adding up the manu-facturer's specified heat output of all the radiators, plus a 3kW allowance for a hot-water cylinder. Ten per cent is added to allow for exceptionally cold weather. The overall calculation is affected by the heat lost through the walls and ceiling, and also by the number of air changes caused by ventilation.

Some plumbers' merchants will make the relevant calculations for you, if you provide them with the dimensions of each room. Alternatively, you can calc-ulate your requirements yourself, using a software package produced for use with a home computer. There are also purpose-made calculators known as Mears wheels, which can be hired, com-plete with instructions, from a supplier of central-heating equipment.

Ideal room temperatures
A central-heating designer and installer normally aims at providing a system that will heat rooms to the temperatures shown below, assuming an outdoor temperature of -1°C (30°F).

ROOM TEMPERATURE	
Living room	21°C (70°F)
Dining room	21°C (70°F)
Kitchen	16°C (60°F)
Hall/landing	18°C (65°F)
Bedroom	16°C (60°F)
Bathroom	23°C (72°F)

Gas-fired boilers

Many gas-fired boilers have pilot lights that burn constantly, in order to ignite the burners whenever heat is required. The burners may be operated manually or by a timer set to switch the heating on and off at selected times. It is also possible to link the boiler to a room thermostat, so that the heating is switched on and off to keep temperatures at the required level throughout the house. Another thermostat, within the boiler itself, prevents the water from overheating.

An increasing number of boilers have electronic ignition. With this system, the pilot is not ignited until the room thermostat demands heat – then, once the boiler reaches the required tempera-ture, valves to the burner and pilot light close, shutting off the fuel supply until heat is next called for.

Oil-fired boilers

Pressure-jet oil-fired boilers are fitted with controls similar to the ones for gas boilers described above. Oil boilers can be floor-standing or wall-mounted. To run oil-fired central heating, you need a large oil-storage tank outside, with easy access for delivery tankers.

Solid-fuel boilers

Solid-fuel boilers are invariably floor-standing and require a conventional flue. Back boilers are small enough to be built into a fireplace.

Instant control of heat isn't possible with a solid-fuel boiler – the rate at which the fuel is burnt is usually con-trolled by a thermostatic damper and sometimes by a fan.

The system must have some means for the heat to escape in the event of the circulation pump failing (otherwise, the water could boil in the appliance and damage it). This is usually arranged by means of a natural-convection circuit (pipe) that leads from the boiler and the heat exchanger in the domestic hot-water cylinder to a radiator situated in the bathroom, where the excess heat can be used to dry wet towels.

If a solid-fuel boiler is to continue burning, it has to be kept stoked – so some models are made with a hopper feed that tops them up automatically. You need a suitable place to store fuel for the boiler; and the residual ash has to be removed regularly.

A boiler that takes its combustion air from within the house and expels fumes through a conventional open flue (see far left) must have access to a perman-ent ventilator fitted in an outside wall. The ventilator has to be of the correct size – as recommended by the boiler manufacturer – and must not contain a fly-screen mesh, which could become blocked. Refer to Building Regulations F1 – 1.8 for specific guidance. A boiler that is starved of air will create carbon monoxide – a lethal invisible gas that has no smell.

A cupboard that houses a balanced-flue room-sealed boiler must be fitted with ventilators at the top and bottom, to prevent the boiler overheating.

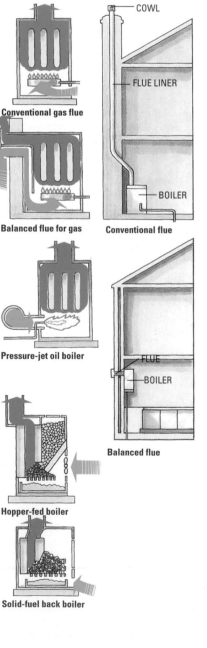

Conventional gas flue

Balanced flue for gas

COWL

FLUE LINER

BOILER

Conventional flue

Pressure-jet oil boiler

FLUE

BOILER

Balanced flue

Hopper-fed boiler

Solid-fuel back boiler

Condensing boilers

Condensing boilers extract more heat from the fuel than other types of boiler. This is achieved either by passing the water through a highly efficient heat exchanger or by having a secondary heat exchanger that uses heat from the flue to 'preheat' cool water returning from the radiators.

With a conventional boiler, the moisture within the exhaust gases passes through the flue as steam. Since a condensing boiler extracts more heat from the gases, much of the moisture they contain condenses within the boiler. The water thus produced is collected at the bottom of the combustion chamber and drained through a small pipe.

Another by-product on a cold damp day is a light cloud of water vapour at the flue outlet, where the relatively cool exhaust gases meet the outside air. This could be a nuisance if the flue is sited close to a neighbour's window. There are regulations governing the siting of balanced flues – check the requirements with your Building Control Officer.

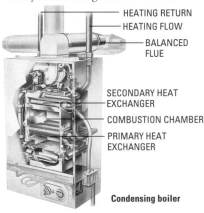

HEATING RETURN
HEATING FLOW
BALANCED FLUE
SECONDARY HEAT EXCHANGER
COMBUSTION CHAMBER
PRIMARY HEAT EXCHANGER

Condensing boiler

Combination boilers

Combination boilers provide both hot water to a sealed heating system and a separate supply of instant hot water directly to taps and showers. The advantages are ease of installation (there are no tanks or pipes in the loft), space-saving (there's no hot-water storage cylinder) and economy (you heat only the water you use).

The main drawback is a fairly slow flow rate – so it takes longer to fill a bath, and it's not usually possible to use two hot taps at the same time. Combination boilers are therefore best suited to small households or flats. However, to overcome these problems, the newer generation of combination boilers incorporate a small built-in hot-water storage tank.

The hot water from a central-heating boiler is pumped along small-bore pipes connected to radiators (or convectors), mounted at strategic points to heat individual rooms and hallways. The standard radiator is a double-skinned pressed-metal panel, which is heated by the hot water that flows through it. Despite its name, a radiator emits only a fraction of its output as radiant heat – the rest being delivered by natural convection as the surrounding air comes into contact with the hot surfaces of the radiator. As the warmed air rises towards the ceiling, cooler air flows in around the radiator, and this air in turn is warmed and moves upwards. As a result, a very gentle circulation of air takes place in the room, and the temperature gradually rises to the optimum set on the room thermostat.

Panel radiators

Radiators are available in a wide range of sizes. The larger they are, the greater their heat output. Output for a given size can be increased further by using 'double radiators', which are made by joining two panels one behind the other. Most types of radiator have fins attached to their rear faces to induce convected heat.

The handwheel valve at one end of the radiator turns the flow of water on or off; the lockshield valve at the other end is set to balance the system, then left alone. An ordinary handwheel valve can be fitted at either end of a radiator, regardless of the direction of flow. However, thermostatic valves, which regulate the temperature of individual radiators, are marked with arrows to indicate the direction of flow and must be fitted accordingly.

A bleed valve, fitted at one of the top corners, is used to release air that has gradually built up inside the radiator. Air trapped inside a radiator prevents the panel from heating properly.

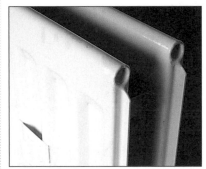

Double-panel radiator

Finned radiator

Decorative radiators
As a rule, flat-panel radiators are designed to be as innocuous as possible. If you prefer something more conspicuous, choose from one of the more colourful ranges. Some radiators are chromed.

Panel radiator
1 A manual handwheel valve turns the flow on or off.
2 A lockshield valve is set to balance the system.
3 A bleed valve disperses airlocks.

Heat emission
As it's heated by the radiator, convected air flows upwards and is replaced by cooler air near the base of the radiator. In addition, heat radiates from the surface of the panel.

☛ **SEE ALSO:** Thermostatic valves 57, Bleeding radiators 61

Radiators and convectors

Convectors

Convector heaters can be used as part of a wet central-heating system. Some models are designed for inconspicuous fixing at skirting level.

Convectors emit none of their heat in the form of direct radiation. The hot water from the boiler passes through a finned pipe inside the heater, and the fins absorb the heat and transfer it to the air around them. The warmed air passes through a damper-controlled vent at the top of the heater, and at the same time cool air is drawn in through the open bottom to be warmed in turn.

With a fan-assisted convector heater, the airflow is accelerated over the fins in order to speed up room heating.

Rising warm air draws in cool air below

Skirting radiators

A skirting radiator is a space-saving alternative to a conventional panel radiator and is designed for install-ation in place of a wooden skirting board. The twin copper-lined water-ways and the outer casing are formed from a single aluminium extrusion.

Made in 6m (19ft 6in) lengths and available in various finishes, skirting radiators are cut to length then joined at the corners of the room, using con-ventional soldered pipe joints. The pipework and valves are hidden from view, but are readily accessible.

An electrically heated version is also available.

Skirting radiator
1 Aluminium extrusion
2 Copper-lined
  waterways
3 Mounting assembly

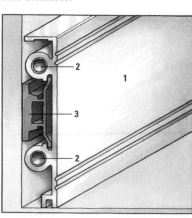

Positioning radiators and convectors

At one time central-heating radiators and convector units were nearly always placed under windows, because the area around a window tends to be the coldest part of a room. However, if you've fitted double glazing to reduce heat loss and draughts, then it would be more efficient to place your heaters elsewhere – especially if your windows are hung with long curtains.

Finned radiators – which accelerate convection considerably – afford a greater degree of flexibility in the siting of heaters and, size permitting, still keep the whole room at a comfortable temperature.

The shape of a room can also affect the siting of heaters and perhaps their number. For example, it is difficult to heat a large L-shaped room with just a single radiator at one end. In situations like this it's probably best to consult a heating installer beforehand, to help you decide upon the optimum number of heaters and their siting.

Wherever possible, avoid hanging curtains or standing furniture in front of a radiator or convector heater. Both curtains and furniture absorb radiated heat – and curtains also tend to trap convected heat behind them.

The warm air rising from a radiator will eventually discolour the paint or wallcovering above it. Fitting a narrow shelf about 50mm (2in) above a radi-ator avoids staining, without inhibiting convection. Alternatively, enclose the radiator in a narrow cabinet – heat output is barely reduced, provided air is able to pass through the enclosure freely, especially at the top and bottom (see below).

Radiator cabinets

Whereas a standard panel radiator may suit a modern interior, it can look out of place in a period-style room. One solution is to enclose the radiator in a cabinet that's more in keeping with the character of the interior.

The cabinet must be ventilated to allow air into the bottom and for the convected warm air to exit from the top. A perforated panel is usually fitted across the front to dissipate the heat and add to the unit's appearance.

Cabinets are available in kit form to fit standard-size radiators. Alternatively, you can cut custom-made panels from MDF board.

Making your own cabinet
A radiator cabinet can be designed to stand on the floor or to be hung on the wall at skirting height. A floor-standing version is described here.

Cut the shelf member (1) and two end panels (2) from 18mm (¾in) MDF.

Make these components large enough to enclose the radiator and both valves. Cut a notch near the base of each end panel to fit the profile of the skirtings.

Glue the panels to the shelf with dowels joints, and dowel a 50 x 25mm (2 x 1in) tie rail (3) between the sides at skirting level. Cut a new skirting moulding (4) to fit along the base of the cabinet, but first cut away the bottom edge of the moulding on the front to form a large vent. Complete the box by applying a decorative moulding (5) around the edge of the shelf.

Cut a front panel (6) from either per-forated hardboard, MDF, aluminium sheet or bamboo lattice, and mount it in a rebated MDF frame (7). Make the frame fit the box, leaving a vent along the top edge. Hold the frame in place with magnetic catches.

Paint the cabinet and, when it is dry, attach it to the wall with metal corner brackets or mirror plates.

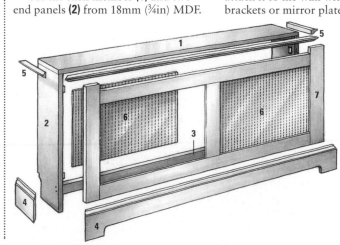

**Floor-standing
radiator cabinet**
1 Shelf
2 End panel
3 Tie rail
4 Skirting
5 Moulding
6 Perforated
  panel
7 Frame

☞ **SEE ALSO:** Radiators 55

Controls for central heating

The various automatic control systems and devices available for wet central heating can, if used properly, provide savings in running costs by reducing wastage of heat to a minimum.

Three basic devices

Automatic controllers can be divided into three basic types: temperature controllers (thermostats), automatic on-off switches (programmers and timers), and heating-circuit controllers (zone valves). These devices can be used, individually or in combination, to provide a very high level of control.

It must be added that they are really effective with gas or oil-fired boilers only, since these can be switched on and off at will. When they're linked to solid-fuel boilers, which take time to react to controls, automatic control systems are much less effective.

ZONE-CONTROL VALVES

There's very little point in heating rooms that aren't being used. In most households, for example, the bedrooms are unoccupied for the greater part of the day and to heat them continuously would be wasteful.

One way of avoiding such waste is to divide your central-heating system into circuits or 'zones' (the usual ones being upstairs and downstairs) and to heat the whole house only when necessary. However, if you divide your house into zones, make sure the unheated areas are adequately ventilated, in order to prevent condensation.

Control is provided by motorized valves linked to a timer or programmer that directs the heated water through selected pipes at predetermined times of day. Alternatively, zone valves linked to individual thermostats can be used to provide separate temperature control for each zone.

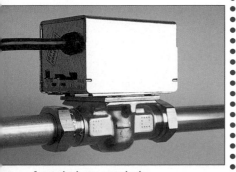

A motorised zone-control valve

Thermostats

All boilers incorporate thermostats to prevent overheating. An oil-fired or gas boiler will have one that can be set to vary heat output by switching the unit on and off; and some models are also fitted with modulating burners, which adjust flame height to suit heating requirements. On a solid-fuel boiler, the thermostat opens and closes a damper that admits more or less air to the firebed to increase or reduce the rate of burning, as required.

A room thermostat – 'roomstat' for short – is often the only form of central-heating control fitted. It is placed in a room where the temperature usually remains fairly stable, and works on the assumption that any rise or drop in the temperature will be matched by similar variations throughout the house.

Roomstats control the temperature by means of simple on-off switching of the boiler – or the pump, if the boiler has to run constantly in order to provide hot water. The main drawback of a roomstat is that it makes no allowance for local temperature changes in other rooms – caused, for example, by the sun shining through a window or a separate heater being switched on.

More sophisticated temperature control is provided by a thermostatic valve, which can be fitted to a radiator instead of the standard manually operated valve. A temperature sensor opens and closes the valve, varying the heat output to maintain the desired temperature in the individual room. Thermostatic radiator valves need not be fitted in every room. You can use one to reduce the heat in a kitchen or small bathroom, for example, while a roomstat regulates the temperature throughout the rest of the house.

The most sophisticated thermostatic controller is a boiler-energy manager or 'optimizer'. This device collects data from sensors inside and outside the building in order to deduce the optimum running period for the central-heating system, so the boiler is not wastefully switched on and off in rapid cycles.

Timers and programmers

You can cut fuel bills substantially by ensuring that the heating is not on while you are out or asleep. A timer can be set so that the system is switched on to warm the house before you get up and goes off just before you leave for work, then comes on again shortly before you return home and goes off at bedtime. The simpler timers provide two 'on' and two 'off' settings, which are normally repeated every day. A manual override enables you to alter the times for weekends and other changes in routine.

More sophisticated devices, known as programmers, offer a larger number of on-off programs – even a different one for each day of the week – as well as control of domestic hot water.

Boiler-energy manager

Room thermostat

Programmer or timer

Thermostatic radiator valve

Heating controls

There are a number of ways to control heating:
1 A wiring centre connects the controls in the system.
2 A programmer/timer is used in conjunction with a zone valve to switch the boiler on or off at pre-set times, and run the heating and hot-water systems.
3 Optional boiler-energy manager controls the efficiency of the heating system.
4 Room thermostats are used to control the pump or zone valves to regulate the overall temperature.
5 A non-electrical thermostatic radiator valve controls the temperature of an individual heater.

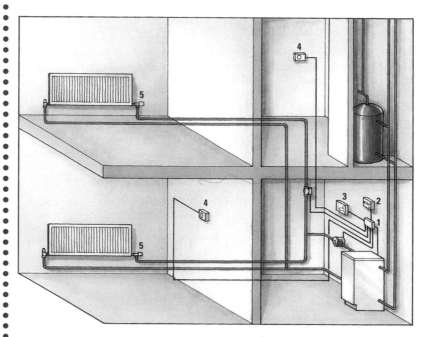

Diagnosing heating problems

When heating systems fail to work properly, they can exhibit all sorts of symptoms, some of which can be difficult to diagnose without specialized knowledge and experience. However, it pays to check out the more commons faults, summarized below, before calling out a heating engineer.

Hissing or banging sounds from boiler or heating pipes

This is caused by overheating due to:

- **Blocked chimney**
  (if you have a solid-fuel boiler).
  Sweep chimney to clear heavy soot.

- **Build-up of scale due to hard water.**
  Shut down boiler and pump. Treat system with a descaler, then drain, flush and refill system.

- **Faulty boiler thermostat.**
  Shut down boiler. Leave pump working to circulate water, to cool system quickly. When it's cool, operate boiler thermostat control. If you don't hear a clicking sound, call in an engineer.

- **Lack of water in system.**
  Shut down boiler. Check feed-and-expansion tank in loft. If empty, the valve may be stuck. Move float-valve arm up and down to restore flow and fill system. If this has no effect, check to see if mains water has been turned off by accident or (in winter) if supply pipe is frozen.

- **Pump not working**
  (with a solid-fuel boiler).
  Shut down boiler, then check that pump is switched on. If pump is not running, turn off power and check wired connections to it. If pump seems to be running but outlet pipe is cool, check for airlock by opening pump bleed screw. If pump is still not working, shut it down, drain system, remove pump and check for blockage. Clean pump or, if need be, replace it.

Radiators in one part of the house do not warm up

- **Timer or thermostat that controls relevant zone valve is not set properly or is faulty.**
  Check timer or thermostat setting and reset if need be. If this has no effect, switch off power supply and check wired connections. If that makes no difference, call in an engineer.

- **Zone valve itself is faulty.**
  Drain system and replace or repair the valve.

- **Pump not working.**
  See above.

All radiators remain cool, though boiler is operating normally

- **Pump not working.**
  Check pump by listening or feeling for motor vibration. If pump is running, check for airlock by opening bleed valve. If this has no effect, the pump outlet may be blocked. Switch off boiler and pump, remove pump and clean or replace as necessary. If pump is not running, switch off and try to free spindle. Look for a large screw in the middle – removing or turning it will reveal the slotted end of the spindle. Turn this until the spindle feels free, then switch pump on again.

- **Pump thermostat or timer is set incorrectly or is faulty.**
  Adjust thermostat or timer setting. If that has no effect, switch off power and check wiring connections. If they are in good order, call in an engineer.

Single radiator doesn't warm up

- **Handwheel valve is closed.**
  Open the valve.

- **Thermostatic radiator valve is set too low or is faulty.**
  Adjust valve setting. If this has no effect, drain the system and replace the valve.

- **Lockshield valve not set properly.**
  Remove lockshield cover and adjust valve setting until radiator seems as warm as those in other rooms. Have lockshield valve properly balanced when the system is next serviced.

- **Radiator valves blocked by corrosion.**
  Close both radiator valves, remove radiator and flush out.

Area at top of radiator stays cool, though bottom is warm

- **Airlock at top of radiator is preventing water circulating fully.**
  Bleed radiator to release trapped air.

Cool patch in centre of radiator, though top and ends are warm

- **Deposits of rust at bottom of radiator are restricting circulation of water.**
  Close both radiator valves, remove radiator and flush out.

Boiler not working

- **Thermostat set too low.**
  Check that roomstat and boiler thermostats are set correctly.

- **Timer or programmer not working.**
  Check that timer or programmer is switched on and set correctly. Have it replaced if fault persists.

- **Gas boiler's pilot light goes out.**
  Relight pilot following instructions supplied with the boiler (these are usually printed on the back of the front panel). If pilot fails to ignite, have it replaced.

Continuous drip from overflow pipe of feed-and-expansion tank in loft

- **Faulty float valve or leaking float, causing valve to stay open.**
  Shut off mains water supply to feed-and-expansion tank and bale it out to below level of float valve. Remove valve and fit new washer. Alternatively, unscrew leaking float from arm and fit new one.

- **Leaking heat-exchanger coil in hot-water cylinder.**
  In this case, dripping from the overflow will occur only if the feed-and-expansion tank is positioned below the cold-water storage tank. Turn off boiler and mains water. Let system cool, then take dip-stick measurement in both tanks. Don't use water overnight – then check again in morning. If the water level has risen in the feed-and-expansion tank and dropped in the cold-water storage tank, have the coil tested.

Water leaking from system

- **Loose pipe unions at joints, pump connections, boiler connections, etc.**
  Turn off boiler (or close down solid-fuel appliance, raking out coals) and switch off pump, then tighten leaking joints. If this has no effect, drain the system and remake joints completely.

- **Split or punctured pipe.**
  Wrap rags around the damaged pipe temporarily, then switch off boiler and pump and make a temporary repair with hose or commercial leak sealant. Drain the system and fit new pipe.

☞ **SEE ALSO:** Draining the system 8, Repairing leaks 9, Float valves 13, Pipework 19–27, Plumbing joints 21–5, Bleed valve 61, Removing/replacing radiators 61, 63

Draining and refilling

Although it's inadvisable to do so unnecessarily, there may be times when you have to drain your wet central-heating system completely and refill it. This could be for routine maintenance, when dealing with a fault, or because you have decided to extend the system or upgrade the boiler. The job can be done fairly easily if you follow the procedures outlined here.

Draining the system

Before draining your central-heating system, cool the water by shutting off the boiler and leaving the circulation pump running. The water in the system will cool quite quickly.

Switch off the pump and turn off the mains water supply to the feed-and-expansion tank in the loft either by closing the stopcock in the feed pipe or by laying a batten across the tank and tying the float arm to it.

The main draincock for the system will normally be in the return pipe near the boiler. Push one end of a garden hose onto its outlet and lead the other end of the hose to a gully or soakaway in the garden, then open the draincock. If you have no key for its square shank, use an adjustable spanner.

Most of the water will drain from the system, but some will be held in the radiators. To release the trapped water, start at the top of the house and carefully open the radiator bleed valves. Air will flow into the tops of the radiators, breaking the vacuum, and the water will drain out. Last of all, drain inverted pipe loops (see below).

Before refilling the system, check that you have closed all the draincocks and radiator bleed valves.

Restore the water supply to the feed-and-expansion tank in the loft. As the system fills up, air will be trapped in the tops of the radiators – so when the water stops running, bleed all the radiators, starting at the bottom of the house. You may also have to bleed the circulating pump. Finally, check all the draincocks and bleed valves for leaks, and tighten them if necessary.

Draincock key
A special tool, similar in principle to a radiator-valve key, is available for operating draincocks.

Tightening a leaking draincock

Cleaning the system

After installing or modifying a central-heating system, flush the pipework with water to get rid of swarf and flux, which can induce corrosion or damage valves or the pump.

To protect the pump during cleaning, it's best to remove it, bridging the gap with a short length of pipe. But it is much easier to turn the pump impeller with a screwdriver before running the system after flushing, in order to make sure it's clear. If you can feel resistance, drain the system and remove the motor, then clean and refit the impeller.

Descaling

If your system is old or badly corroded, a harsh cleaner or descaler may expose minor leaks sealed by corrosion – so use a mild cleanser, introduced into the system via the feed-and-expansion tank or inject it into a radiator via the bleed valve. Manufacturers' instructions vary, but in principle run the cleanser through the system for a week, with the boiler set to a fairly high temperature.

Afterwards, turn off and drain the system, then refill and drain it several times – if possible, using a hose to run mains-pressure water through the system while draining it. Some cleansers must be neutralized before you can add a corrosion inhibitor.

If your boiler is making loud banging noises, treat it and the immediate pipework with a fairly powerful descaler, running the hot-water program only.

Draining procedure
Turn off the mains supply to the tank at the feed-pipe stopcock (1). If there's no stopcock, tie the float-valve arm to a batten laid across the tank (2). With a hose pushed onto the main draincock (3) and its other end at a gully or soakaway outside, open the draincock and let the system empty. Release any water trapped in the radiators (4) by opening their bleed valves (5), starting at the top of the house. Be sure to close all draincocks before you refill the system.

● **Power-flushing the system**
After upgrading an older system, perhaps with a new boiler or radiators, you could flush the system yourself (see left), but it's advisable to have it cleansed thoroughly by a heating engineer, using a power-flushing unit. When it is connected, the unit pumps chemically-treated water through the system to flush out impurities.

Inverted pipe loops

Often when fitting a central-heating system in a house that has a solid ground floor, installers run the heating pipes from the boiler into the ceiling void and drop them down the walls to the individual radiators. Each of these 'inverted pipe loops' has its own draincock. When you're draining the system, they must be drained separately after the main system has been emptied.

RADIATOR

DRAINCOCK

INVERTED PIPE LOOP

BOILER

An inverted pipe loop has its own draincock

☞ SEE ALSO: Turning off the water 6–9, Gully 17, Bleeding radiators 61, Bleeding a pump 64

Maintaining your boiler

The efficiency of modern oil-fired and gas boilers depends on their being checked and serviced annually. Because the mechanisms involved are so complex, the work must be done by a qualified engineer. With either type of boiler, you can enter into a contract for regular maintenance with your fuel supplier or the original installer.

● **Servicing gas boilers**
Any maintenance that involves dismantling any part of a gas boiler must be carried out by a CORGI-registered engineer, who should undertake all the necessary gas-safety checks as part of the service. There's no point in attempting to service the boiler yourself if you are not qualified and equipped to do so – it can also be dangerous, and you will be breaking the law.

Corrosion in the system

Modern boilers and radiators are made from fairly thin materials, and if you fail to take basic anti-corrosion measures, the life of the system can be reduced to 10 years or less. Corrosion may result either from hard-water deposits or from a chemical reaction between the water and the system's metal components.

Lime scale
Scale builds up quickly in hard-water areas of the country. Even a thin layer of lime scale on the inner wall of a boiler's heat exchanger reduces its efficiency and may cause banging and dog-like howling within the system. In fact, the scale can insulate sections of the heat exchanger to such an extent that it produces 'hot spots', leading to premature failure of the component.

Rust
Rust corrodes steel components, most notably radiators. Most rusting occurs within weeks of filling the system; but if air is being sucked in constantly, then rusting is progressive. Having to bleed radiators regularly is a sure sign that air is being drawn into the system.

Sludge
Magnetite (black sludge) clogs the pump and builds up in the bottom of radiators, reducing their heat output.

Electrolytic action
Dissimilar metals, such as copper and aluminium, act like a battery in the acidic water that is present in some central-heating systems. This results in corrosion.

Reducing corrosion

Drain about half a litre (1 pint) of water from the boiler or a radiator. Orange water denotes rusting, and black the presence of sludge. In either case, treat immediately with corrosion inhibitor.

If there are no obvious signs of corrosion, compare the sample with tap water. Drop two plain steel nails into a screw-top glass jar containing some of the sample water, and place two similar nails in a jar of clean tap water. After a couple of days the nails in the tap water should rust; but if your heating system contains sufficient corrosion inhibitor, the nails in the sample jar will remain bright. If they show signs of corrosion, your system needs topping up with inhibitor. It is important to use the same product that is already present in the system – if you don't know what that is, drain and flush the system, then refill with fresh water and inhibitor.

If the test proves inconclusive, check the sample jar after a month or so: if the nails have begun to rust, then the inhibitor needs topping up.

Adding corrosion inhibitor
You can slow down corrosion by adding a proprietary corrosion inhibitor to the water. This is best done when the system is first installed – but the inhibitor can be introduced into the system at any time, provided the boiler is descaled

before doing so. If the system has been running for some time, it is better to flush it out first by draining and refilling it repeatedly until the water runs clean. Otherwise, drain off about 20 litres (4 gallons) of water – enough to empty the feed-and-expansion tank and a small amount of pipework – then pour the inhibitor into the tank and restore the water supply, which will carry the inhibitor into the pipes. About 5 litres (1 gallon) will be enough for most systems, but check the manufacturer's instructions. Finally, switch on the pump to distribute the inhibitor throughout the system.

Reducing scale
You can buy low-voltage coils to create a magnetic field that will prevent the heat exchanger of your boiler becoming coated with scale. However, unless you have soft water in your area, the only way to actually avoid hard water in the system is to install a water softener.

Phosphate balls are sometimes used to prevent the formation of scale in an instantaneous boiler. But unless the dispenser is regulated to release just the right amount, there's a danger of overdosing the system with phosphates.

Before fitting any device to reduce scale, it is essential to seek the boiler manufacturer's advice.

Locating gas boilers
Modern boilers fit snugly into standard kitchen cupboards.

It pays to have your central-heating system serviced regularly. Check the Yellow Pages for a suitable engineer, or ask the original installer of the system if he or she is willing to undertake the necessary servicing.

Gas installations

Gas suppliers offer a choice of servicing schemes for boilers. These are primarily provided to cover the suppliers' own installations, but they will also service systems put in by other installers if a satisfactory inspection of the installation by the supplier is carried out first.

The simplest of the schemes provides for an annual check and adjustment of the boiler. If any repairs are found to be necessary, either at the time of the regular check or at other times during the year, then the labour and necessary parts are charged separately. But for an extra fee it is possible to have both free labour and free parts for boiler repairs at any time of year. The gas supplier will also extend the arrangement to include inspection of the whole heating system when the boiler is being checked, plus free parts and labour for repairs to the system.

You may find that your installer or a local firm of CORGI heating engineers offers a similar choice of servicing and maintenance contracts. The best course is to compare the schemes and decide which gives greatest value for money.

Oil-fired installations

Both installers of oil-fired central-heating systems and suppliers of fuel oil offer servicing and maintenance contracts similar to those outlined above for gas-fired systems. The choice of schemes available ranges from a simple annual check-up to complete cover for parts and labour whenever repairs are necessary.

As with the schemes for gas, it pays to shop around and make a comparison of the various services on offer and the charges that apply.

Solid-fuel systems

If you have a solid-fuel system, it is important to keep the chimney and the flueway swept. The job, which should be done twice a year, is very similar to sweeping an open-fire chimney, access being either through the front of a room heater that has a back boiler or through a soot door in the flue pipe or chimney breast.

When you have swept the chimney, clean out the boiler with a stiff brush and remove the dust and soot with a vacuum cleaner. Lift out any broken fire bars and drop new ones in place.

☞ **SEE ALSO:** Water softeners 48, Switching off electricity 68

Bleeding the system

There are a number of reasons why it may be necessary to remove a radiator – for example, to make decorating the wall behind it easier. You can remove individual radiators without having to drain the whole system.

Make sure you have plenty of rag to hand for mopping up spilled water, plus a jug and a large bowl. The water in the radiator will be very dirty – so, if possible, roll back the floorcovering before you start.

Shut off both valves, turning the shank of the lockshield valve clockwise with a key or an adjustable spanner **(1)**. Note the number of turns needed to close it, so that later you can reopen it by the same amount.

Unscrew the cap-nut that keeps the handwheel valve or lockshield valve attached to the adaptor in the end of the radiator **(2)**. Hold the jug under the joint and open the bleed valve slowly to let the water drain out. Transfer the water from the jug to the bowl, and continue doing this until no more water can be drained off.

Unscrew the cap-nut that keeps the other valve attached to the radiator, lift the radiator free from its wall brackets, and drain any remaining water into the bowl **(3)**. If you're going to decorate the wall, unscrew the brackets.

To replace the radiator, screw the brackets back in place, then rehang the radiator and tighten the cap-nuts on both valves. Close the bleed valve and reopen both radiator valves (open the lockshield valve by the same number of turns you used when closing it). Last of all, bleed the air from the radiator.

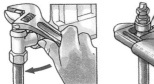

1 Close the valve   **2 Unscrew cap-nut**

3 Final draining
Lift radiator from brackets and drain off any remaining water.

Trapped air prevents radiators heating up fully, and regular intake of air can cause corrosion. If a radiator feels cooler at the top than at the bottom, it's likely that a pocket of air has formed inside it and is impeding full circulation of the water. Getting the air out of a radiator – 'bleeding' it – is a simple procedure.

Bleeding a radiator

First switch off the circulation pump – and preferably turn off the boiler too, although that is not vital.

Each radiator has a bleed valve at one of its top corners, identifiable by a square-section shank in the centre of the round blanking plug. You should have been given a key to fit these shanks by the installer; but if not, or if you have inherited an old system, you can buy a key for bleeding radiators at any DIY shop or ironmonger's.

Use the key to turn the valve's shank anticlockwise about a quarter of a turn. It shouldn't be necessary to turn it further – but have a small container handy to catch spurting water, in case you open the valve too far. You will probably also need some rags to mop up water that dribbles from the valve. Don't try to speed up the process by opening the valve further than necessary to let the air out – that is likely to produce a deluge of water.

You will hear a hissing sound as the air escapes. Keep the key on the shank of the valve; then when the hissing stops and the first dribble of water appears, close the valve tightly.

Blocked bleed valve

If no water or air comes out when you attempt to bleed a radiator, check whether the feed-and-expansion tank in the loft is empty. If the tank is full of water, then the bleed valve is probably blocked with paint.

Close the inlet and outlet valve at each end of the radiator, then remove the screw from the centre of the bleed valve. Clear the hole with a piece of wire, and reopen one of the radiator valves slightly to eject some water from the hole. Close the radiator valve again and refit the screw in the bleed valve. Open both radiator valves and test the bleed valve again.

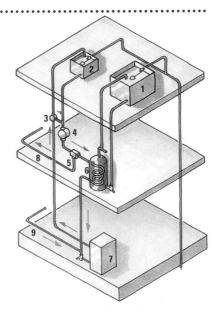

Dispersing an air pocket in a radiator

Fitting an air separator

If you find you are having to bleed a radiator or radiators frequently, a large quantity of air is entering the system. This situation should be remedied before it leads to serious corrosion.

Check that the feed-and-expansion tank in the loft is not acting like a radiator and warming up when you run the central heating or hot water. This would indicate that hot water is being pumped through the vent pipe into the tank and taking air with it back into the system. To cure the problem, fit an air separator in the vent pipe and link it to the cold feed that runs from the feed-and-expansion tank.

If the pump is fitted on the return pipe to the boiler, it may be sucking in air through the unions or even through leaking spindles on radiator valves.

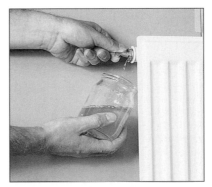

Heating system with air separator
1 Cold-water storage tank
2 Feed-and-expansion tank
3 Air separator
4 Pump
5 Motorized valve
6 Hot-water cylinder
7 Boiler
8 Radiator flow
9 Radiator return

Replacing radiator valves

Like taps, radiator valves can develop leaks – which are usually relatively easy to cure. Occasionally, however, it's necessary to replace a faulty valve.

Curing a leaking radiator valve

VALVE HEAD

GLAND NUT

Leaking spindle
To stop a leak from a radiator-valve spindle, tighten the gland nut with a spanner. If the leak persists, undo the nut and wind a few turns of PTFE tape down into the spindle.

Water leaking from a radiator valve is probably seeping from around the spindle (see left). However, when the water runs round and drips from the valve's cap-nut, it's the nut that often appears to be the source of the leak. Dry the valve, then hold a paper tissue against the various parts of the valve to ascertain exactly where the moisture is coming from. If the nut is leaking, tighten it gently; if that's unsuccessful, undo and reseal it (see left).

Grip leaky valve with wrench and tighten cap-nut

Replacing a worn or damaged valve

● **Resealing a cap-nut**
Drain the system and undo the leaking nut. Smear the olive with silicone sealant and retighten the cap-nut. Don't overtighten the nut or you may damage the olive. As an alternative to sealant, wind two turns of PTFE tape around the olive (not around the threads).

To replace a radiator valve, first drain the system, then lay rags under the valve to catch the dregs. Holding the body of the valve with a wrench (or water-pump pliers), use an adjustable spanner to unscrew the cap-nuts that hold the valve to the pipe (1) and also to the adaptor in the end of the radiator. Lift the valve from the end of the pipe (2); if you're replacing a lockshield valve, be sure to close it first – counting the turns, so you can open the new valve by the same number to balance the radiator.

Unscrew the valve adaptor from the radiator (3). You may be able to use an adjustable spanner, depending on the type of adaptor, or may find you need a hexagonal radiator spanner.

Fitting the new valve
Ensure that the threads in the end of the radiator are clean. Drag the teeth of a hacksaw across the threads of the new adaptor to roughen them slightly, then wind PTFE tape four or five times round them. Screw the adaptor into the end of the radiator and tighten with a spanner.

Slide the valve cap-nut and a new olive over the end of the pipe and fit the valve (4) – but don't tighten the cap-nut yet. First, holding the valve body with a wrench, align it with the adaptor and tighten the cap-nut that holds them together (5). Then tighten the cap-nut that holds the valve to the water pipe (6). Refill the system and check for leaks.

The spindle of a Belmont valve is sealed with O-rings – which you can replace without having to drain the radiator.

To find out which O-rings you need, take the plastic head of the valve to a plumbers' merchant before you begin work. On very old valves the rings are green, whereas the newer rings are red.

Wrap an old towel around the valve body and undo the spindle (which has a left-hand thread). A small amount of water will leak out at first – but as you continue to remove the spindle, water pressure seals the valve automatically.

Two O-rings are housed in grooves in the spindle. Prise off the rings, using the tip of a small screwdriver, and then lubricate the spindle with a smear of silicone grease. Slide the new rings into position and replace the spindle.

O-rings are housed in grooves in the valve spindle

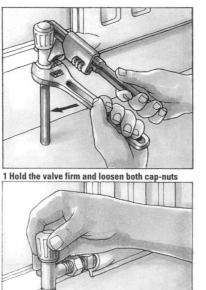

1 Hold the valve firm and loosen both cap-nuts

2 Unscrew the cap-nuts and lift the valve out

3 Remove the valve adaptor from the radiator

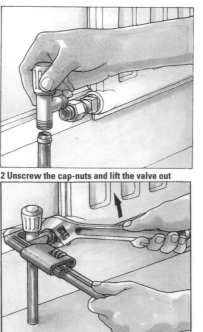

4 Fit new adaptor, then fit the new valve on the pipe

5 Connect valve to adaptor and tighten cap-nut

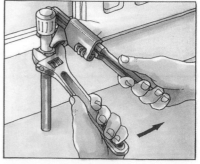

6 Tighten cap-nut that holds the valve to the pipe

☛ **SEE ALSO:** Draining the system 59, Adjustable spanner 77, Pipe wrench 78, PTFE tape 81

Replacing a radiator

Try to obtain a new radiator exactly the same size as the one you're planning to replace. This makes the job relatively easy.

Simple replacement

Drain the old radiator and remove it from the wall. Then unscrew the two valve adaptors at the bottom of the radiator, using an adjustable spanner or a hexagonal radiator spanner. Next, use a bleed key to unscrew the bleed valve; then remove both of the blanking plugs from the top of the radiator, using a radiator spanner **(1)**.

Clean any corrosion from the threads of the adaptors and blanking plugs with wire wool **(2)**, then wind four or five turns of PTFE tape round the threads **(3)**. Screw the plugs and adaptors into the new radiator; and then screw the bleed valve into its blanking plug.

Hang the new radiator on the wall brackets and connect the valves to their adaptors. Open the valves, then fill and bleed the radiator.

1 Removing the plugs
Use a radiator spanner to unscrew the two blanking plugs at the top of the radiator.

2 Cleaning the threads
Use wire wool to clean any corrosion from the threads of the blanking plugs and valve adaptors.

3 Taping the threads
Make the threaded joints watertight by wrapping four or five turns of PTFE tape round the plugs and adaptors before you screw them into the new radiator. Use a hacksaw blade to roughen the threads, in order to encourage the tape to grip.

Installing a different-pattern radiator

More work is involved in replacing a radiator if you can't get another one of the same pattern. You will probably have to fit new wall brackets and alter the pipe runs.

Drain your central-heating system, then take the old brackets off the wall. Lay the new radiator face down on the floor and slide one of its brackets onto the hangers welded to the back of the radiator. Measure the position of the brackets and transfer these measurements to the wall **(1)**. You need to allow a clearance of 100 to 125mm (4 to 5in) below the radiator.

Line up the new radiator brackets with the pencil marks on the wall, and mark the fixing-screw holes for them.

Drill and plug the holes, then screw the brackets in place **(2)**.

Take up the floorboards below the radiator and sever the vertical portions of the feed and return pipes (either cap the old T-joints or replace them with straight joints). Connect the valves to the bottom of the radiator and hang it on its brackets.

Slip a new vertical pipe into each of the valves and, using either capillary or compression fittings, connect these pipes to the original pipework running under the floor **(3)**. Tighten the nuts connecting the new pipes to the valves.

Finally, refill the system with water, and check all the new connections and joints for leaks.

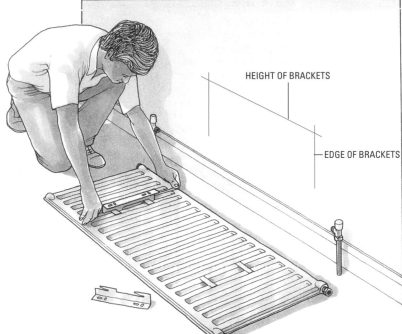

HEIGHT OF BRACKETS

EDGE OF BRACKETS

1 Transferring the measurements
Measure the positions of the radiator brackets and transfer these dimensions to the wall.

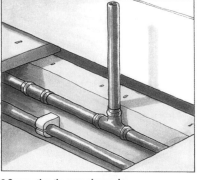

3 Connecting the new pipework.
Make sure the vertical section of pipe aligns with the radiator valve.

2 Securing the brackets
Screw the mounting brackets to the wall.

☛ **SEE ALSO:** **Connecting pipes 20–3, 25–7, Draining the system 59, Bleeding radiators 61, Removing radiators 61, Adjustable spanner 77, Radiator spanner 77, PTFE tape 81**

Servicing a pump

Wet central heating depends on a steady cycle of hot water pumped from the boiler to the radiators then back to the boiler for reheating. If the pump is not working properly, the result is poor circulation or none at all. Adjusting or bleeding the pump may be the answer; otherwise, it may need replacing.

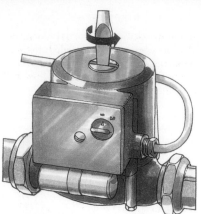

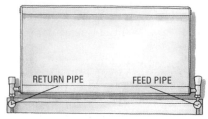

Open the bleed valve with a screwdriver

Bleeding the pump

If an airlock forms in the circulation pump, the impeller spins ineffectually and your radiators fail to warm up properly. The cure is to bleed the air from the pump, a procedure similar to bleeding a radiator. Have a jar handy to catch any spilled water.

Look for a screw-in bleed valve in the pump's outer casing. Then switch off the pump and open the bleed valve slightly with a screwdriver or vent key until you hear air hissing out. When the hissing stops and a drop of water appears, close the bleed valve.

Using an infra-red thermometer
This is a relatively sophisticated – and costly – thermometer for measuring the temperature drop across a radiator. To obtain an instant reading, simply aim the sensor at the pipe just below the radiator valve.

Adjusting the pump

Basically, there are two types of central-heating pump: fixed-head and variable-head. Fixed-head pumps run at a single speed, forcing the heated water round the system at a fixed rate. The speed of variable-head pumps is adjustable.

When fitting a variable-head pump, the installer balances the radiators, then adjusts the pump's speed to achieve an optimum temperature for every room. If you can't boost a room's temperature by opening the radiator's handwheel valve, try adjusting the pump speed. However, before adjusting the pump, you should check that all your radiators show the same temperature drop between their inlets and outlets. To test your radiators, you can obtain a pair of clip-on thermometers from a plumbers' merchant.

Clip one of the thermometers to the feed pipe just below the radiator valve; and the other one to the return pipe, also below its valve (**1**). The difference between the temperatures registered by the thermometers should be about 11°C (20°F). If it's not, close the lock-shield valve slightly to increase the difference in temperature; or open the valve to reduce it.

Having balanced all the radiators, you can now adjust the pump's speed by one increment at a time (**2**) until the radiators are giving the overall temperatures you require. Depending on the make and model of pump, you may need to use a special tool, such as an Allen key, to make the adjustments. Switch off the pump before making each adjustment.

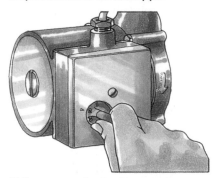

RETURN PIPE   FEED PIPE

1 Clip thermometers to the radiator pipes

2 Adjust pump speed to alter the temperature

● **Bridging the gap**
Modern pumps are sometimes smaller than equivalent older models. If this proves to be the case, buy a converter designed to bridge the gap in the existing pipework.

Replacing a worn pump

If you have to replace a faulty pump, make sure you buy a new one that is equivalent in performance. If in doubt, consult a professional installer.

First, turn off the boiler and close the isolating valves situated on each side of the pump. If the pump lacks isolating valves, you will have to drain down the whole system.

At your consumer unit, identify the electrical circuit that supplies the pump and remove the relevant circuit fuse or MCB. Then take the coverplate off the pump (**1**) and disconnect its wiring.

With a bowl or bucket ready to catch the water from the pump, undo the nuts that hold the pump to the valves or pipework (**2**).

Having removed the old pump, install the new one (**3**), taking care to fit correctly any sealing washers that are provided. Tighten the connecting nuts.

Remove the coverplate from the new pump and feed in the flex. Connect the wires to the pump's terminals (**4**), then replace the coverplate. If the pump is of the variable-head type (see above), set the speed control to match the speed indicated on the old pump.

Open both isolating valves – or refill the system, if you had to drain it – then check the pump connections for leaks.

Open the pump's bleed valve to release any trapped air. Finally, replace the fuse or MCB in the consumer unit and test the pump.

1 Remove coverplate   2 Undo connecting nuts

3 Attach new pump   4 Connect power flex

☞ **SEE ALSO:** Draining the system 59, Filling the system 59, Removing a fuse 72

If a motorised valve ceases to open, its electric motor may have failed. Before replacing the motor, use a mains tester to check whether it's receiving power. If it is, fit a new motor.

There is no need to drain the system. Switch off the electricity supply to the central-heating system (see right)— don't merely turn off the programmer, as motorized valves have a permanent live feed.

Once the power is off, remove the cover and undo the single screw that holds the motor in place **(1)**. Open the valve, using the manual lever, and lift out the motor **(2)**. Disconnect the two motor wires by cutting off the connectors.

Insert a new motor – available from a plumbers' merchant – then let the lever spring back to the closed position. Fit and tighten the retaining screw. Strip the ends and connect the wires, using the new connectors supplied **(3)**.

Replace the valve cover, and test the operation by turning on the power and running the system.

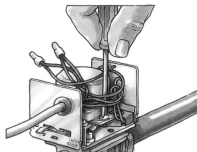

1 Releasing the motor-retaining screw
Remove the cover and then the retaining screw.

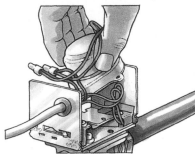

2 Removing the motor
Push the lever to open the valve, then lift out motor.

3 Fitting the new motor
Join the wires, using the two connectors supplied.

Replacing a control valve

Control valves are the means by which timers and thermostats adjust the level of heating. Worn or faulty control valves can seriously impair the reliability of the system, and should therefore be repaired or replaced promptly.

Replacing a faulty valve

When you buy a new valve, make sure it is of exactly the same pattern as the one you are replacing.

Drain the system. Then, at your consumer unit, remove the fuse or MCB for the circuit to which the central-heating controls are connected.

The flex from the valve will be wired to an adjacent junction box, which is also connected to the heating system's other controls. Take the cover off the box and disconnect the wiring for the valve – making a note of the terminals used, to make reconnection easier.

To remove the old valve, simply cut through the pipe on each side **(1)**. When fitting the new valve, bridge the gap with short sections of pipe, complete with joints at each end **(2)**. Spring the assembly into place and connect the joints to the old pipe, then tighten the valve cap-nuts **(3)**. Connect the valve's flex to the junction box, then insert the circuit fuse or MCB.

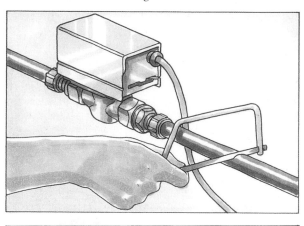

1 Removing the valve
If you're unable to disconnect the valve, use a hacksaw to cut through the pipe on each side.

2 Fitting the new valve
With the new valve connected to short sections of pipe, spring the assembly into the pipe run.

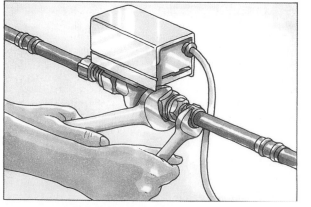

3 Tightening the nuts
Having connected the pipes, tighten the valve cap-nuts on each side, using a pair of spanners. Refill the heating system and check that the valve is working properly.

Two-port control valve
A two-port valve seals off a section of pipework when the water has reached the required temperature.

Three-port control valve
This type of valve can isolate the central heating from the hot-water circuit.

Compression coupling

Soldered coupling

Slip couplings
It can sometimes be difficult to replace a valve using two conventional joints. If you can't spring the new assembly into place (see left), use a slip coupling at one end. This coupling is free to slide along the pipe to bridge the gap.

☞ **SEE ALSO:** Making pipe joints 20, Heating controls 57, Draining the system 59, Removing a fuse 72

Underfloor heating

With the availability of flexible plastic plumbing, sophisticated controls and efficient insulation, underfloor heating has become a viable and affordable form of central heating. Specialist manufacturers have developed a range of warm-water heating systems to suit virtually any situation. The same companies generally offer a design service aimed at providing a heating system that satisfies the customer's specific requirements. An installation manual is delivered along with the necessary materials and equipment.

BENEFITS OF UNDERFLOOR HEATING

Although it's easier to incorporate underfloor heating while a house is being built, installing it in an existing building is by no means impossible. And there's no reason why underfloor heating can't be made to work alongside a panel-radiator system – it could provide the ideal solution for heating a new extension or conservatory, for example.

Compared with panel radiators, an underfloor-heating system radiates heat more evenly and over a wider area. This has the effect of reducing hot and cold spots within the room and produces a more comfortable environment, where the air is warmest at floor level and cools as it rises towards the ceiling.

Underfloor heating is also energy-efficient, because it operates at a lower temperature than other central-heating systems – and because there's a more even temperature throughout a room, the roomstat can be set a degree or two lower, yet the house still feels warm and cosy. The net result is a saving on fuel costs and, with relatively cool water in the return cycle, a modern condensing boiler works even more efficiently.

Because there are no radiators or convectors to accommodate, you have greater freedom when planning the layout of furnishings. The floors can be finished with any conventional covering, but the thermal resistance of the flooring needs to be taken into account when the system is designed.

● **Combining systems**
You can have radiators upstairs and underfloor heating downstairs. A mixing manifold will allow you to combine the two systems, using the same boiler. Any type of boiler is suitable for underfloor heating, but a condensing boiler is the most economic.

● **Maintaining underfloor heating**
The heating elements are virtually maintenance free. If the flow through the pipework becomes restricted, then the circuit can be flushed through with mains-pressure water by attaching a hose to the manifold.

Underfloor-heating systems

Underfloor heating can be incorporated in any type of floor construction, including solid-concrete floors, boarded floating floors and suspended timber floors (see below). The heat emanates from a continuous length of plastic tube that snakes across the floor, forming parallel loops and covering an area of one or more rooms.

The entire floor area is divided into separate zones to provide the most efficient layout. Each zone is controlled by a roomstat and is connected to a thermostatically controlled multi-valve manifold that forms the heart of the system. The manifold controls the temperature of the water and the flow rate to the various zones. Once a room or zone reaches its required temperature, a valve automatically shuts off that part of the circuit. A flow meter for each of the zones allows the circuits to be balanced when setting up the system and subsequently monitors its performance.

The manifold, which is installed above floor level, is connected to the boiler via a conventional circulation pump.

Methods for installing underfloor heating

When underfloor heating is installed in a new building, the plastic tubes are usually set into a solid-concrete floor **(1)**. Flooring insulation is laid over the base concrete, and rows of special pipe clips are fixed to the insulation; sometimes a metal mesh is used instead of the clips. The flexible heating tubes are then clipped into place at the required spacings (see opposite), and a concrete screed is poured on top.

With a boarded floating floor **(2)**, a layer of grooved insulation is laid over the concrete base, and the pipes are set in aluminium 'diffusion' plates inserted in the grooves. The entire floor area is then covered with an edge-bonded chipboard or a similar decking material.

The heating pipes can be fastened with spacer clips to the underside of a suspended wooden floor **(3)**. In this situation, clearance holes are drilled through the joists at strategic points to permit a continuous run of pipework. Reflective foil and thick blanket insulation are then fixed below the pipes.

It is possible to lay the pipes on top of a suspended floor, but this method raises the floor level by the thickness of the pipe assembly and the new decking.

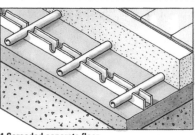

1 Screeded concrete floor

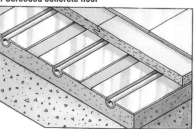

2 Boarded floating floor

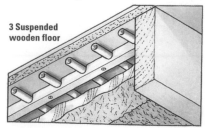

3 Suspended wooden floor

☞ **SEE ALSO:** Panel radiators 55, Thermostats 57

Installing underfloor heating

Added to an existing radiator system, underfloor heating makes a good choice for heating a new conservatory extension. The large areas of glass in a conservatory present very few options for placing radiators, and the concrete slab that is typically used for conservatory floors provides an ideal base for this form of heating.

WHERE TO START

Send the details of your proposed extension to the underfloor-heating supplier. The company will also need a scaled plan of your house and the basic details of your present central-heating system in order to be able to supply you with a well-planned scheme and quotation.

You can expect to receive a complete package, including all the components and an installation manual.

Your options

The simplest type of system will be connected to the pipework of your existing radiator circuit. Heat for the extension will only be available when the existing central heating is running, although the temperature in the conservatory can be controlled independently by a roomstat connected to a motorised zone valve and the underfloor-heating pump.

For full control, the flow and return pipework to the underfloor system must be connected directly to the boiler, and the roomstat must be wired up to switch the boiler on and off and to control the temperature of the conservatory.

If it proves impossible to utilize the existing heating system, or the boiler has insufficient capacity and cannot be upgraded, then you would need to have an independent boiler and pump system to heat the conservatory.

The basic plumbing system

Your supplier will suggest the best point to connect your new plumbing to the existing central-heating circuit. It can be at any convenient point, provided that the performance of your radiators will not be affected.

The pipework connecting the manifold for the underfloor heating to the radiator circuit can be metal or plastic, and it can be the same size as but not larger than the existing pipes. Again your supplier will advise what to use.

The flow and return pipes from the manifold to the conservatory circuit (illustrated here, as an example) are connected to individual zone distributors, which in turn are connected to the flexible underfloor-heating tubes.

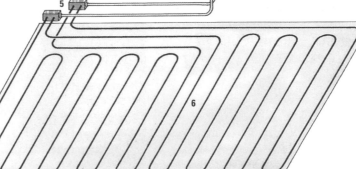

Basic system
1 Flow and return pipes from existing central-heating circuit.
2 Water-temperature mixing valve.
3 Pump
4 Manifold with zone valves.
5 Zone distributors.
6 Underfloor-heating tube.

Floor construction
1 Blinded hardcore
2 DPM
3 Concrete base
4 Insulation
5 Edge insulation
6 Pipe clips and pipe
7 Screed
8 Floor tiles

Constructing the floor

You will need to excavate the site and lay a concrete base as recommended by the conservatory manufacturer, a surveyor, or your local Building Control Officer (BCO). The base must include a damp-proof membrane (DPM). Allow for a covering layer of floor insulation – a minimum of 50mm (2in) flooring-grade expanded polystyrene or 30mm (1¼in) extruded polyurethane (check with your BCO). The floor should be finished with a 65mm (2½in) sand-and-cement screed, plus the preferred floorcovering.

When laying the floor insulation slabs, you should install a strip of insulation, 25mm (1in) thick, all round the edges. This is to prevent cold bridging the masonry walls and the floor screed.

Cut a hole through the house wall, ready for the new plumbing.

Installing the system

Mount the manifold in a convenient place and connect the two distributor blocks below it – one for the flow, and the other for the return. Run the flow and return pipes back into the house, ready for connecting to the existing central-heating circuit. Install your new pump and a mixing valve in the flow and return pipes.

Following the layout supplied by the system's manufacturer, press the spikes of the pipe clips into the insulation at the prescribed spacing (1). Lay out the heating tubes for both coils, and clip them into place.

Push the end of one of the coils into the flow distributor, and the other end of the same coil into the return distributor

(2). Connect the other coil similarly.

Connect the flow and return pipes to the house's central-heating system – it pays to insert a pair of isolating valves at this point, so that you can shut off the new circuit for servicing. Fill, flush out and check the new system for leaks.

Apply the screed composed of 4 parts sharp sand : 1 part cement, with a plasticizer additive. Leave it to dry for at least three weeks before laying your floor-covering – don't use heat to accelerate the drying.

Fit the roomstat at head height, out of direct sunlight. Make the electrical connections, then set the roomstat to control the circuit pump and zone valve, following the instructions supplied.

1 Press the pipe clips into place

2 Push tubing into the distributors

☞ **SEE ALSO:** Pipe joints 20, Heating controls 57, Draining the system 59

Main switch equipment

Electricity flows because of a difference in 'pressure' between the live wire and the neutral one, and this difference in pressure is measured in volts. Domestic electricity in this country is supplied as alternating current, at 230 volts, by way of the electricity company's main service cable. This normally enters your house underground, although in some areas electricity is distributed by overhead cables.

The service head

The main cable terminates at the service head, or 'cutout', which contains the service fuse. This fuse prevents the neighbourhood's supply being affected if there should be a serious fault in the circuitry of your house. Cables connect the cutout to the meter, which registers how much electricity you consume. Both the meter and cutout belong to the electricity company and must not be tampered with. The meter is sealed in order to disclose interference.

If you use cheap night-time power for storage heaters and hot water, a time switch will be mounted between the cutout and the meter.

Consumer unit

Electricity is fed to and from the consumer unit by 'meter leads', thick single-core insulated-and-sheathed cables made up of several wires twisted together. The consumer unit is a box that contains the fuseways that protect the individual circuits in the house. It also incorporates the main isolating switch, which you operate when you need to cut off the supply of power to the whole house.

In a house where several new circuits have been installed over the years, the number of circuits may exceed the number of fuseways in the consumer unit. If so, an individual switchfuse unit – or more than one – may have been mounted alongside the main unit. Switchfuse units comprise a single fuseway and an isolating switch; they, too, are connected to the meter by means of meter leads.

If your home is heated by off-peak storage heaters, then you will have an Economy 7 meter and a separate consumer unit for the heater circuits.

● **Cross-bonding cable sizes**
Single-core cables are used to cross-bond gas and water pipes to earth. An electrician can calculate the minimum size for these cables, but for any single house or flat, it is safe to use 10mm² cable. (See also PME opposite).

● **The main isolating switch**
Not all main isolating switches operate the same way. Before you need to use it, check to see whether the main switch on your consumer unit has to be in the up or down position for 'off'.

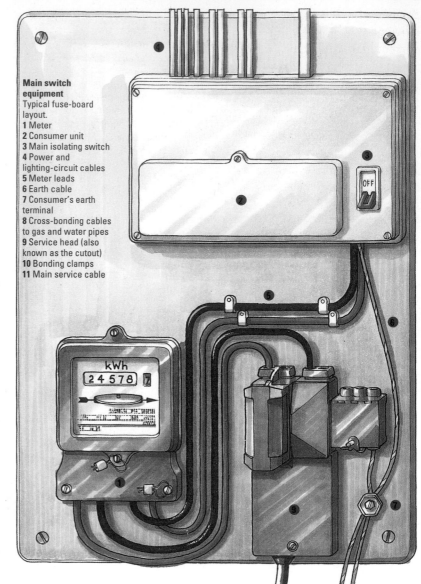

Main switch equipment
Typical fuse-board layout.
1 Meter
2 Consumer unit
3 Main isolating switch
4 Power and lighting-circuit cables
5 Meter leads
6 Earth cable
7 Consumer's earth terminal
8 Cross-bonding cables to gas and water pipes
9 Service head (also known as the cutout)
10 Bonding clamps
11 Main service cable

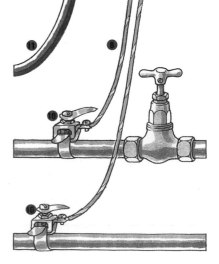

SWITCHING OFF THE POWER

In an emergency, switch off the supply of electricity to the entire house by operating the main isolating switch on the consumer unit.

Before working on any part of the electrical system of your home, always operate the main isolating switch, then remove the individual circuit fuse or miniature circuit breaker (MCB) that will cut off the power to the relevant circuit. That circuit will then be safe to work on, even if you restore the power to the rest of the house by operating the main switch again.

☛ **SEE ALSO:** Switchfuse unit 72, Cheaper electricity 73, Electric shock treatment 80, Circuit breakers 81

Bathroom safety

Because water is such a highly efficient conductor of electric current, water and electricity form a very dangerous combination. For this reason, in terms of electricity bathrooms are potentially the most dangerous areas in your home. Where there are so many exposed metal pipes and fittings, combined with wet conditions, regulations must be stringently observed if fatal accidents are to be avoided.

GENERAL SAFETY

- Sockets must not be fitted in a bathroom – except for special shaver sockets that conform to BS EN 60742 Chapter 2, Section 1.

- The IEE Wiring Regulations stipulate that light switches in bathrooms must be outside zones 0 to 3 (see opposite). The best way to comply with this requirement is to fit only ceiling-mounted pull-cord switches.

- Any bathroom heater must comply with the IEE Wiring Regulations.

- If you have a shower in a bedroom, it must be not less than 3m (9ft 11in) from any socket outlet, which must be protected by a 30 milliamp RCD.

- Light fittings must be well out of reach and shielded – so fit a close-mounted ceiling light, properly enclosed, rather than a pendant fitting.

- Never use a portable fire or other electrical appliance, such as a hairdryer, in a bathroom – even if it is plugged into a socket outside the room.

Supplementary bonding

In any bathroom there are many non-electrical metallic components, such as metal baths and basins, supply pipes to bath and basin taps, metal waste pipes, radiators, central-heating pipework and so on – all of which could cause an accident during the time it would take for an electrical fault to blow a fuse or operate a miniature circuit breaker (MCB). To ensure that no dangerous voltages are created between metal parts, the Wiring Regulations stipulate that all these metal components must be connected one to another by a conductor which is itself connected to a terminal on the earthing block in the consumer unit. This is known as supplementary bonding and is required for all bathrooms – even when there is no electrical equipment installed in the room, and even though the water and gas pipes are bonded to the consumer's earth terminal near the consumer unit.

When electrical equipment such as a heater or shower is fitted in a bathroom, that too must be supplementary-bonded by connecting its metalwork – such as the casing – to the nonelectrical metal pipework, even though the appliance is connected to the earthing conductor in the supply cable.

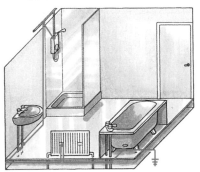

Supplementary bonding in a bathroom

Making the connections

The Wiring Regulations specify the minimum size of earthing conductor that can be used for supplementary bonding in different situations, so that large-scale electrical installations can be costed economically. In a domestic environment, use 6mm^2 single-core cable insulated with green-and-yellow PVC for supplementary bonding. This is large enough to be safe in any domestic situation. For a neat appearance, plan the route of the bonding cable to run from point to point behind the bath panel, under floorboards, and through basin pedestals. If necessary, run the cable through a hollow wall or under plaster, like any other electrical cable.

Connecting to pipework

An earth clamp (1) is used for making connections to pipework. Clean the pipe locally with wire wool to make a good connection between the pipe and clamp, and scrape or strip an area of paintwork if the pipe has been painted.

1 Fit an earth clamp to pipework

Connecting to a bath or basin

Metal baths or basins are made with an earth tag. Connect the earth cable by trapping the bared end of the conductor under a nut and bolt with metal washers (2). Make sure the tag has not been painted or enamelled.

If an old metal bath or basin has not been provided with an earth tag, drill a hole through the foot of the bath or through the rim at the back of the basin; and connect the cable with a similar nut and bolt, with metal washers.

2 Connect to bath or basin earth tag

Connecting to an appliance

Simply connect the earth cable to the terminal provided in the electrical appliance (3) and run it to a clamp on a metal supply pipe nearby.

3 Fix to the earth terminal in an appliance
The appliance's own earth connection may share the same terminal.

WARNING

Have supplementary bonding tested by a qualified electrician. If you have not had any previous experience of wiring and making electrical connections, have supplementary bonding installed by a professional.

Wiring regulations exist to ensure that bathrooms are safe places. However, you should familiarize yourself with what to do if someone does receive an electric shock. *See* p. 68 for further details.

☛ **SEE ALSO:** Bonding to earth 6, Electric shower 38, 41, Switches 70, Bathroom heaters 71, Shaver sockets 71, Electric shock treatment 80, PME 81

69

Zones for bathrooms : UNDER THE BATH

Within a room containing a bath or shower, the IEE Wiring Regulations define areas, or zones, where specific safety precautions apply. The regulations also describe what type of electrical appliances can be installed in each zone, and the routes cables must take in order to serve those appliances. There are special considerations for extra-low-voltage equipment with separated earth; this is best left to a qualified electrician.

The space under a bathtub is designated as zone 1 if it is accessible without having to use a tool – that is, if there is no bath panel or if the panel is attached with magnetic catches or similar devices that allow the panel to be detached without using a tool of some kind. If, however, the panel is screw-fixed – so that it can only be removed with the aid of a screwdriver – then the enclosed space beneath the bath is considered to be outside all zones.

The four zones

Any room containing a bathtub or shower is divided into four zones. Zone 0 is the interior of the bathtub or shower tray – not including the space beneath the tub, which is covered by other regulations (see top right). Zones 1 to 3 are specific areas above and all round the bath or shower, where only specified electrical appliances and their cables may be installed. Wiring outside these areas must conform to the IEE Wiring Regulations, but no specific 'zone' regulations apply.

ZONE	LOCATION	PERMITTED
Zone 0	Interior of the bathtub or shower tray.	No electrical installation.
Zone 1	Directly above the bathtub or shower tray, up to a height of 2.25m (7ft 5in) from the floor. (See also top right.)	Instantaneous water heater. Instantaneous shower. All-in-one power shower, with a suitably waterproofed integral pump. The wiring that serves appliances within the zone.
Zone 2	Area within 0.6m (2ft) horizontally from the bathtub or shower tray in any direction, up to a height of 2.25m (7ft 5in) from the floor. The area above zone 1, up to a height of 3m (9ft 11in) from the floor.	Appliances permitted in zone 1. Light fittings. Extractor fan. Space heater. Whirlpool unit for the bathtub. Shaver socket to BS EN 60742 Chapter 2, Section 1. The wiring that serves appliances within the zone and any appliances in zone 1.
Zone 3	Up to 2.4m (7ft 11in) outside zone 2, up to a height of 2.25m (7ft 5in) from the floor. The area above zone 2 next to the bathtub or shower, up to a height of 3m (9ft 11in) from the floor.	Appliances permitted in zones 1 and 2. Any fixed electrical appliance (a heated towel rail, for example) that is protected by a 30 milliamp RCD. The wiring that serves appliances within the zone and any appliances in zones 1 and 2.

Supplementary bonding

In bathrooms, non-electrical metallic components must be bonded to earth (see opposite). In zones 1, 2 and 3, this supplementary bonding is required to all pipes, any electrical appliances and any exposed metallic structural components of the building. This does not include window frames, unless they are themselves connected to metallic structural components.

Supplementary bonding is not required outside the zones. And in the special case of a bedroom containing a shower cubicle, supplementary bonding can also be omitted from zone 3.

Switches

Electrical switches, including ceiling-mounted switches operated by a pull cord, must be situated outside the zones. The only exceptions are those switches and controls incorporated in appliances suitable for use in the zones.

If the bathroom ceiling is higher than 3m (9ft 11in), ceiling-mounted pull-cord switches can be mounted anywhere. However, if the ceiling height is between 2.25 and 3m (7ft 5in and 9ft 11in), pull-cord switches must be mounted at least 0.6m (2ft) – measured horizontally – from the bathtub or shower cubicle. If the ceiling is lower than 2.25m (7ft 5in), switches can only be mounted outside the room.

IP coding

Electrical appliances installed in zones 1 and 2 must be manufactured with suitable protection against splashed water. This is designated by the code IPX4 (the letter X is sometimes replaced with a single digit). Any number larger than four is also acceptable as this indicates a higher degree of water-proofing. If in doubt, check with your supplier that the appliance is suitable for its intended location.

● **Cable runs**
You are not permitted to run electrical cables that are feeding a zone through another zone designated with a lower number. This includes cables buried in the plaster or concealed behind other wallcoverings.

● **13amp sockets**
In the special case of a bedroom containing a shower cubicle, socket outlets are permitted in the room, but only outside the zones, and the circuit that feeds the sockets must be protected by a 30 milliamp RCD.

IP coding
Suitable equipment may be marked with the symbol shown above.

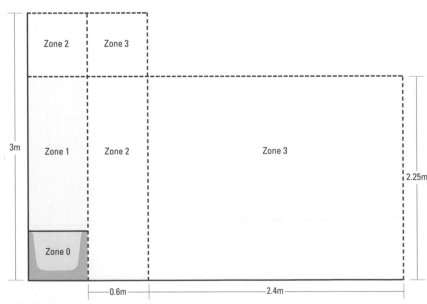

3m

Zone 2 | Zone 3

Zone 1 | Zone 2 | Zone 3

2.25m

Zone 0

0.6m | 2.4m

Zones within a room containing a bath or shower

☞ **SEE ALSO:** Wiring heaters 71, Wiring a shower unit 71, Electric shock treatment 80

Wiring heaters

An electrically heated shower unit is plumbed into the mains water supply. The flow of water operates a switch to energize an element that heats the water on its way to the shower spray-head. Because there's so little time to heat the flowing water, instantaneous showers use a heavy load – from 6 to 10.8kW. Consequently, an electrically heated shower unit has to have a separate radial circuit, which must be protected by a 30 milliamp RCD.

The circuit cable needs to be 10mm² two-core-and-earth. For showers up to 10.3kW, the circuit should be protected by a 45amp MCB or fuse, either in a spare fuseway at the consumer unit or in a separate single-way consumer unit fitted with a 30 milliamp RCD. A 10.8kW shower needs a 50amp MCB. The cable runs directly to the shower unit, where it must be wired according to the manufacturer's instructions.

The shower unit itself has its own on/off switch, but there must also be a separate isolating switch in the circuit. This must not be accessible to anyone using the shower, so you need to install a ceiling-mounted 45amp double-pole pull-switch (a 50amp switch is required for a 10.8kW shower). The switch has to be fitted with an indicator that tells you when the switch is 'on'. Fix the backplate of the switch to the ceiling and, having sheathed the earth wires with a green-and-yellow sleeve, connect them to the E terminal on the switch. Connect the conductors from the consumer unit to the switch's 'Mains' terminal, and those of the cable to the shower to the 'Load' terminals **(1)**.

The shower unit and all metal pipes and fittings must be bonded to earth.

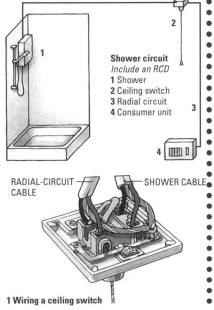

Shower circuit
Include an RCD
1 Shower
2 Ceiling switch
3 Radial circuit
4 Consumer unit

RADIAL-CIRCUIT CABLE   SHOWER CABLE

1 Wiring a ceiling switch

When you're installing a skirting heater or wall-mounted heater or an oil-filled radiator, wire the appliance to a fused connection unit mounted nearby, at a height of about 150 to 300mm (6in to 1ft) from the floor. Whether the connection to the unit is by flex or cable will depend on the type of appliance. Follow the manufacturer's instructions for wiring, and fit the appropriate fuse in the connection unit.

In a bathroom, a fused connection unit must be mounted outside zones 0 to 3. Any heater that is mounted near the floor of a bathroom must therefore be wired to a connection unit installed outside the room. If the appliance is fitted with flex, mount a flexible-cord outlet **(1)** next to the appliance – and then run a cable from the outlet to the fused connection unit outside the bathroom and connect it to the 'Load' terminals in the unit.

The flexible-cord outlet is mounted either on a standard surface-mounted box or flush on a metal box. At the back of the faceplate are three pairs of terminals to take the conductors from the flex and the cable **(2)**.

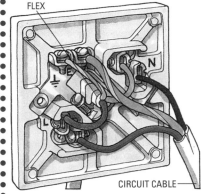

FLEX

N

L

CIRCUIT CABLE

2 Wiring a flexible-cord outlet

Shaver sockets

Special shaver socket outlets are the only kind of electrical socket allowed in bathrooms. They contain a transformer that isolates the user side of the unit from the mains, reducing the risk of an electric shock.

This type of socket has to conform to the exacting British Standard BS EN 60742. However, there are shaver sockets that do not have an isolating transformer and therefore don't conform to this standard. These are quite safe to install and use in a bedroom – but this type of socket must not be fitted in a bathroom.

Radiant wall heaters

Radiant wall heaters for use in bathrooms must be fixed high on the wall, outside zones 0 to 2. A fused connection unit fitted with a 13amp fuse (or 5amp fuse for a heater of 1kW or less) must be mounted at a high level outside the zones, and the heater must be controlled by a double-pole pull-cord switch (with this type of switch, both live and neutral contacts are broken when it is off). Many heaters have a built-in double-pole switch; otherwise, you must fit a ceiling-mounted 15amp double-pole switch between the fused connection unit and the heater. Switch terminals marked 'Mains' are for the cable on the circuit side of the switch; those marked 'Load' are for the heater side. The earth wires are connected to a common terminal on the switch box.

If it is not possible to run a spur to the fused connection unit from a socket outside the bathroom, run a separate radial circuit from the connection unit to a 15amp fuseway in the consumer unit, using 2.5mm² cable. In either case, the circuit should be protected by a 30 milliamp RCD.

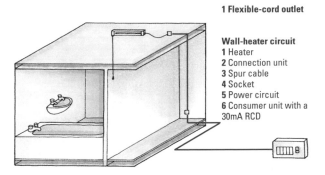

1 Flexible-cord outlet

Wall-heater circuit
1 Heater
2 Connection unit
3 Spur cable
4 Socket
5 Power circuit
6 Consumer unit with a 30mA RCD

You can wire a shaver socket from a junction box on an earthed lighting circuit or from a fused connection unit, fitted with a 3amp fuse, on a ring-circuit spur. If you're installing the shaver socket in a bathroom, then the fused connection unit must be positioned outside the room. Run 1mm² two-core-and-earth cable from the connection unit to the shaver socket; then connect the conductors: red to L and black to N **(1)**. Sheath the earth wire with a green-and-yellow sleeve and connect it to E.

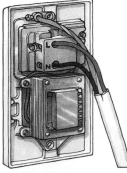

1 Wiring a shaver unit

★ **SEE ALSO: Switching off power 68, Bathroom safety 69, Zones for bathrooms 70, Switches 70, Electric shock treatment 80**

71

Fused connection units

A fused connection unit is a device for joining the flex (or cable) of an appliance to circuit wiring. The connection unit incorporates the added protection of a cartridge fuse.

Changing a fuse
With the electricity turned off, remove the retaining screw in the face of the fuse holder. Take the holder from the connection unit; prise out the old fuse and fit a new one; then replace the holder and the retaining screw.

If the appliance is connected by a flex, choose a unit that has a cord outlet in the faceplate.

Some fused connection units are fitted with a switch, and some of these have a neon indicator that shows at a glance whether they are switched on. A switched connection unit allows you to isolate the appliance from the mains.

All fused connection units are single (there are no double versions available) with square faceplates that fit metal boxes for flush mounting or standard surface-mounted plastic boxes.

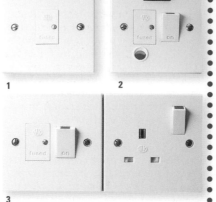

Fused connection units
1 Unswitched connection unit.
2 Switched unit with cord outlet and indicator.
3 Connection unit and socket outlet in a dual mounting box.

Wiring a fused connection unit

Fused connection units can be supplied by a ring circuit, a radial circuit or a spur. Some appliances are connected to the unit with flex, others with cable. Either way, the wiring arrangement inside the unit is the same. Units with cord outlets have clamps to secure the connecting flex.

An unswitched connection unit has two live (L) terminals – one marked 'Load' for the brown wire of the flex, and the other marked 'Mains' for the red wire from the circuit cable. The blue wire from the flex and the black wire from the circuit cable go to similar neutral (N) terminals; and both earth wires are connected to the unit's earth (E) terminal or terminals **(1)**.

Switched connection unit
A fused connection unit with a switch also has two sets of terminals. Those marked 'Mains' are for the spur or ring cable that supplies the power; the terminals marked 'Load' are for the flex or cable from the appliance.

Wire up the flex side first, connecting the brown wire to the L terminal, and the blue one to the N terminal, both on the 'Load' side. Connect the green-and-yellow wire to the E terminal **(2)** and tighten the cord clamp.

Attach the circuit conductors to the 'Mains' terminals – red to L, and black to N; then sleeve the earth wire and take it to the E terminal **(2)**.

If the fused connection unit is on a ring circuit, you must fit two circuit conductors into each 'Mains' terminal and into the earth terminal. Before securing the unit in its box with the fixing screws, make sure the wires are held firmly in the terminals and can fold away neatly.

The water in a storage cylinder can be heated by an electric immersion heater, providing a central supply of hot water for the whole house. The heating element is rather like a larger version of the one that heats an electric kettle. It is normally sheathed in copper, but more expensive sheathings of incoloy or titanium will increase the life of the element in hard-water areas.

Adjusting the water temperature
The thermostat that controls the maximum temperature of the water is set by adjusting a screw inside the plastic cap covering the terminal box **(1)**.

Types of immersion heater
An immersion heater can be installed either from the top of the cylinder or from the side, and top-entry units can have single or double elements.

With the single-element top-entry type, the element extends down almost to the bottom of the cylinder, so that all of the water is heated whenever the heater is switched on **(2)**.

For economy, one of the elements in the double-element type is a short one for daytime top-up heating, while the other is a full-length element that heats the entire contents of the cylinder, using the cheaper night-rate electricity **(3)**. A double-element heater that has a single thermostat is called a twin-element heater; one with a thermostat for each element is known as a dual-element heater.

Side-entry elements are of identical length. One is positioned near to the bottom of the cylinder, and the other a little above half way **(4)**.

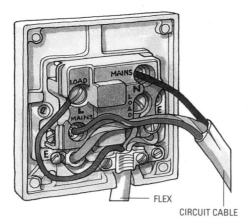

1 Wiring a fused connection unit

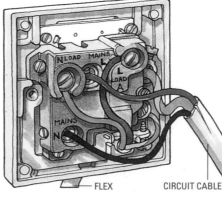

2 Wiring a switched fused connection unit

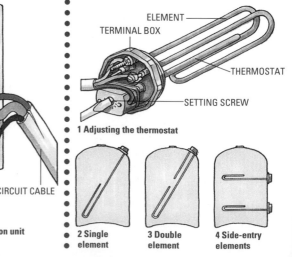

ELEMENT
TERMINAL BOX
THERMOSTAT
SETTING SCREW

1 Adjusting the thermostat

2 Single element **3 Double element** **4 Side-entry elements**

★ **SEE ALSO:** Switching off power 68, Circuits 79, Electric shock treatment 80

Wiring immersion heaters

If you agree to their installing a special meter, your electricity company will supply you with cheap-rate power for seven hours sometime between midnight and 8.00 a.m., the exact period being at the discretion of the company. This scheme is called Economy 7.

Provided you have a cylinder that is large enough to store hot water for a day's requirements, you can benefit by heating all your water during the Economy 7 hours. Even if you heat your water electrically only in summer, the scheme may be worthwhile. For the water to retain its heat all day, you must have an efficient insulating jacket fitted to the cylinder or a cylinder already factory-insulated with a layer of heat-retaining foam.

If your cylinder is already fitted with an immersion heater, you can use the existing wiring by fitting an Economy 7 programmer, a device that will switch your immersion heater on automatically at night and heat up the whole cylinder. Then if you occasionally run out of hot water during the day, you can always adjust the programmer's controls to boost the temperature briefly, using the more expensive daytime rate.

You can make even greater savings if you have two side-entry immersion heaters or a dual-element one. The programmer will switch on the longer element, or the bottom one, at night; but if the water needs heating during the day, then the upper or shorter element is used.

Economy 7 without a programmer

You can have a similar arrangement without a programmer if you wire two separate circuits for the elements. The upper element is wired to the daytime supply, while the lower one is wired to its own switchfuse unit and operated by the Economy 7 time switch during the hours of the night-time tariff only. A setting of 75°C (167°F) is recommended for the lower element, and 60°C (140°F) for the upper one. If your water is soft or your heater elements are sheathed in titanium or incoloy, you can raise the temperatures to 80°C (175°F) and 65°C (150°F) respectively without reducing the life of the elements.

To ensure that you never run short of hot water, leave the upper unit switched on permanently. It will only start heating up if the thermostat detects a temperature of 60°C (140°F) or less, which should happen very rarely if you have a large cylinder that is properly insulated.

The circuit

The majority of immersion heaters are rated at 3kW; but although you can wire most 3kW appliances to a ring circuit, an immersion heater is regarded as using 3kW continuously, even though rarely switched on all the time. A continuous 3kW load would seriously reduce a ring circuit's capacity, so immersion heaters must have their own radial circuits.

The circuit needs to be run in 2.5mm² two-core-and-earth cable protected by a 15amp fuse. Each element must have a double-pole isolating switch mounted near the cylinder; the switch should be marked 'WATER HEATER' and have a neon indicator (**1**). A 2.5mm² heat-resistant flexible cord runs from the switch to the immersion heater.

If the cylinder is situated in a bathroom, the switch must be outside zones 0 to 2. If this precludes an ordinary water-heater switch, fit a 20amp ceiling-mounted pull-switch with a mechanical ON/OFF indicator.

Wiring side-entry heaters

For simplicity use two switches, one for each heater and marked accordingly.

Wiring the switches
Fix the two mounting boxes to the wall, feed a circuit cable to each, and wire them in the same way. Strip and prepare the wires, then connect them to the 'Mains' terminals – red to L, black to N. Sheath the earth wire in a green-and-yellow sleeve and fix it to the common earth terminal (**2**).

Prepare a heat-resistant flex for each switch. At each one, connect the green-and-yellow earth wire to the common earth terminal and the other wires to the 'Load' terminals – brown to L, and blue to N (**2**). Then tighten the flex clamps and screw on the faceplates.

Wiring the heaters
The flex from the upper switch goes to the top heater, and the flex from the lower switch to the bottom one. At each heater, feed the flex through the hole in the cap and prepare the wires.

Connect the brown wire to one of the terminals on the thermostat (the other one is already connected to the wire running to an L terminal on the heating element). Connect the blue wire to the N terminal, and the green-and-yellow wire to the E terminal (**3**). Then replace the caps on the terminal boxes.

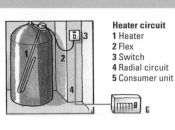

Heater circuit
1 Heater
2 Flex
3 Switch
4 Radial circuit
5 Consumer unit

CIRCUIT CABLE

FLEX TO HEATER

2 Wiring the switch

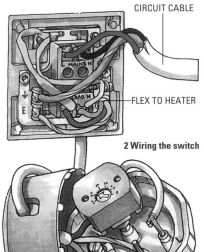

3 Wiring the heater

Running the cable
Run the circuit cables from the cylinder cupboard to the fuse board; then, with the power switched off, connect the cable from the upper heater to a spare fuseway in the consumer unit. Although the consumer unit is switched off, the cable between the main switch and the meter will remain live – so take special care. Wire the other cable to its own switchfuse unit – or to your storage-heater consumer unit, if you have one – ready for connection to the Economy 7 time switch. Make the connections as described for a cooker circuit.

Dual-element heaters

Wire the immersion-heater circuit as described above, but feed the flex from both switches into the cap on the heater. Connect the brown wire from the upper switch to the L2 terminal on the one thermostat, and the other brown wire to the L1 terminal on the second thermostat (**4**). Connect the blue wires to their respective neutral terminals (**4**). Connect both earth wires to the E terminal.

1 RCD protection
When installing any electrical appliance in a bathroom, the circuit should be protected by a 30 milliamp RCD.

1 A 20amp switch for an immersion heater

LIVE
L2

EARTH
N1   E   N2
NEUTRAL

4 Make sure your heater is fitted with two thermostats, as shown.

★ **SEE ALSO:** Switching off power 68, Consumer units 68, Zones for bathrooms 70, Circuit lengths 79

Plumbing tools

PLUMBER'S AND METALWORKER'S TOOL KIT

Although plastics have been used for drainage for some time, the advent of ones suitable for mains-pressure and hot water has affected the plumbing trade more radically. However, brass fittings and pipework made from copper and other metals are still extensively used for domestic plumbing, so the plumber's tool kit is still basically for working metal.

EQUIPMENT FOR REMOVING BLOCKAGES

You don't have to get a plumber to clear blocked appliances, pipes or even main drains. All the necessary equipment can be bought or hired.

Sink plunger

This is a simple but effective tool for clearing a blockage from a sink, washbasin or bath trap. A pumping action on the rubber cup forces air and water along the pipe to disperse the blockage. When you buy a plunger, make sure the cup is large enough to cover the waste outlet.

It is possible to hire larger plungers for clearing blockages from WC traps.

● **Essential tools**
Sink plunger
Scriber
Centre punch
Steel rule
Try square
General-purpose
hacksaw

Hydraulic pump

A blocked waste pipe can be cleared with a hand-operated hydraulic pump. A downward stroke creates a powerful jet of water that should push the obstruction clear. If, however, the blockage is lodged firmly, an upward stroke may create enough suction to pull the obstruction out of place.

WC auger

The short coiled-wire WC auger designed for clearing WC and gully traps is rotated by a handle in a rigid, hollow shaft. The auger has a vinyl guard to prevent the WC pan getting scratched.

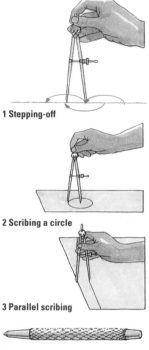

Drain auger

A flexible coiled-wire drain auger will pass through small-diameter waste pipes to clear blockages. Pass the corkscrew-like head into the waste pipe till it reaches the blockage, clamp the cranked handle onto the other end, and then turn it to rotate the head and engage the blockage. Push and pull the auger till the pipe is clear.

SET OF RODS

PLUNGER   CORKSCREW   SCRAPER

Drain rods

You can hire a complete set of rods and fittings for clearing main drains and inspection chambers. The rods come in 1m (3ft 3in) lengths of poly-propylene with threaded brass connectors.

The clearing heads comprise a double-worm corkscrew fitting, a 100mm (4in) rubber plunger and a hinged scraper for clearing the open channels in inspection chambers.

MEASURING AND MARKING TOOLS

Tools for measuring and marking metal are very similar to those used for wood, but they are made and calibrated for greater accuracy because metal parts must fit with precision.

Scriber

For precise work, use a pointed hardened-steel scriber to mark lines and hole centres on metal. Use a pencil to mark the centre of a bend, as a scored line made with a scriber may open up when the metal is stretched on the outside of the bend.

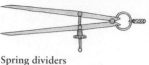

Spring dividers

Spring dividers are similar to a pencil compass, but both legs have steel points. These are adjusted to the required spacing by a knurled nut on a threaded rod that links the legs.

Using spring dividers

Use dividers to step-off divisions along a line (1) or to scribe circles (2). By running one point against the edge of a workpiece, you can scribe a line parallel with the edge (3).

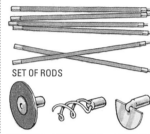

1 Stepping-off

2 Scribing a circle

3 Parallel scribing

Centre punch

A centre punch is an inexpensive tool for marking the centres of holes to be drilled.

Using a centre punch

With its point on dead centre, strike the punch with a hammer. If the mark is not accurate, angle the punch towards the true centre, tap it to extend the mark in that direction, and then mark the centre again.

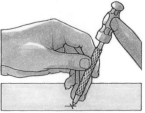

Correcting a misplaced centre mark

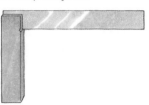

Steel rule

You will need a long tape measure for estimating pipe runs and positioning appliances, but use a 300 or 600mm (1 or 2ft) steel rule for marking out components when absolute accuracy is required.

Try square

You can use a woodworker's try square to mark out or check right angles; however, an all-metal engineer's try square is precision-made for metalwork. The small notch between blade and stock allows the tool to fit properly against a right-angled workpiece even when the corner is burred by filing. For general-purpose work, choose a 150mm (6in) try square.

METAL-CUTTING TOOLS

You can cut solid bar, sheet and tubular metal with an ordinary hacksaw, but there are tools specifically designed for cutting sheet metal and pipes.

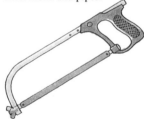

General-purpose hacksaw

A modern hacksaw has a tubular-steel frame with a light cast-metal handle. The frame is adjustable to accommodate replaceable blades of different lengths, which are tensioned by tightening a wing nut.

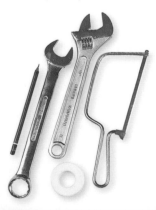

☛ **SEE ALSO: Clearing a WC 17, Clearing drains 18**

Plumbing tools

CHOOSING HACKSAW BLADES

You can buy 200, 250 and 300mm (8, 10 and 12in) hacksaw blades. Try the different lengths till you find the one that suits you best. Choose the hardness and size of teeth according to the type of metal you are planning to cut.

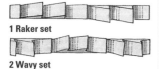

1 Raker set

2 Wavy set

Size and set of teeth

A coarse hacksaw blade has 14 to 18 teeth per 25mm (1in); a fine one has 24 to 32. The teeth are set (bent sideways) to make a cut wider than the blade's thickness, to prevent it jamming in the work. Coarse teeth are 'raker set' **(1)** – with pairs of teeth bent to opposite sides and separated by a tooth left in line with the blade to clear metal waste from the kerf (cut). Fine teeth are too small to be raker set, and the whole row is 'wavy set' **(2)**. Use a coarse blade for cutting soft metals like brass and aluminium, which would clog fine teeth; and a fine blade for thin sheet and the harder metals.

Hardness

A hacksaw blade must be harder than the metal it is cutting, or its teeth will quickly blunt. A flexible blade with hardened teeth will cut most metals, but there are fully hardened blades that stay sharp longer and are less prone to losing teeth. However, being rigid and brittle, they break easily. Blades of high-speed steel are expensive and even more brittle than the fully hardened ones, but they will cut very hard alloys.

Fitting a hacksaw blade

With its teeth pointing away from the handle, slip a new blade onto the pins at each end of the hacksaw frame. Apply tension with the wing nut. If the new blade tends to wander off line as you cut, tighten the wing nut.

Turning a blade

Sometimes it's easier to work with the blade at right angles to the frame. To do so, rotate the square-section spigots a quarter turn before fitting the blade.

1 Turn first kerf away from you

Sawing metal bar

*Hold the work in an engineer's vice, with the marked cutting line as close to the jaws as possible. Start the cut on the waste side of the line with short strokes until the kerf is about 1mm (¹⁄₁₆ in) deep; then turn the bar 90 degrees in the vice, so that the kerf faces away from you, and cut a similar kerf in the new face **(1)**. Continue in this way until the kerf runs right round the bar, then cut through the bar with long steady strokes. Steady the end of the saw with your free hand, and put a little light oil on the blade if necessary.*

Sawing rod or pipe

As you cut a cylindrical rod or tube, rotate it away from you till the kerf runs right round the rod or tube before you sever it.

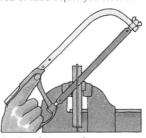

Sawing sheet metal

To saw a small piece of sheet metal, sandwich it between two strips of wood clamped in a vice. Adjust the metal to place the cutting line close to the strips, then saw down the waste side with steady strokes and the blade angled to the work. To cut a thin sheet of metal, clamp it between two pieces of plywood and cut through all three layers simultaneously.

Sawing a groove

To cut a slot or groove wider than a standard hacksaw blade, fit two or more identical blades in the frame at the same time.

Junior hacksaw

Use a junior hacksaw for cutting small-bore tubing and thin metal rod. The simplest ones have a solid spring-steel frame that holds the blade under tension.

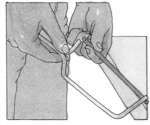

Fitting a new blade

To fit a blade, locate it in the slot at the front of the frame and bow the frame against a workbench until the blade fits in the rear slot.

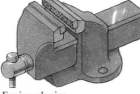

Engineer's vice

A large engineer's or metalworker's vice has to be bolted to the workbench, but smaller ones can be clamped on. Slip soft fibre liners over the jaws of a vice to protect workpieces held in it.

Cold chisel

Plumbers use cold chisels for hacking old pipes out of masonry. They are also useful for chopping the heads off rivets and cutting metal rod. Sharpen the tip of the chisel on a bench grinder.

Straight snips

Universal snips

Tinsnips

Tinsnips are used for cutting sheet metal. **Straight snips** have wide blades for cutting straight edges. If you try to cut curves with them, the waste usually gets caught against the blades; but it is possible to cut a convex curve by progressively removing small straight pieces of waste down to the marked line. **Universal snips** have thick narrow blades that cut a curve in one pass and will also make straight cuts.

Using tinsnips

As you cut along the marked line, let the waste curl away below the sheet. To cut thick sheet metal, clamp one handle of the snips in a vice, so you can apply your full weight to the other one. Try not to close the jaws completely every time, as that can cause a jagged edge on the metal. Wear thick gloves when cutting sheet metal.

SHARPENING SNIPS

- Clamp one handle in a vice and sharpen the cutting edge with a smooth file. File the other edge and finish by removing the burrs from the backs of the blades on an oiled slipstone.

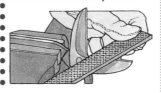

Sheet-metal cutter

Tinsnips tend to distort a narrow strip cut from the edge of a metal sheet. However, the strip remains perfectly flat when removed with a sheet-metal cutter. The same tool is also suited to cutting rigid plastic sheet, which cracks if it is distorted by tinsnips.

Tube cutter

A tube cutter slices the ends off pipes at exactly 90 degrees to their length. The pipe is clamped between the cutting wheel and an adjustable slide with two rollers, and is cut as the tool is moved round it. The adjusting screw is tightened between each revolution.

A pipe slice, which works like a tube cutter, can be operated in confined spaces.

Chain-link cutter

Cut large-diameter pipes with a chain-link cutter. Wrap the chain round the pipe, locate the end link in the clamp, and tighten the adjuster until the cutter on each link bites into the metal. Work the handle back and forth to score the pipe, and continue tightening the adjuster intermittently until the pipe is severed.

● **Essential tools**
Junior hacksaw
Cold chisel
Tinsnips
Tube cutter

Sheet-metal cutter

Tube cutter

Pipe slice

Chain-link cutter

☛ **SEE ALSO:** Cutting pipe 21

Plumbing tools

DRILLS AND PUNCHES

Special-quality steel bits are made for drilling holes in metal. Cut 12 to 25mm (½ to 1in) holes in sheet metal with a hole punch.

Twist drills

Metal-cutting twist drills are similar to the ones used for wood but they are made from high-speed steel and their tips are ground to a shallower angle. Use them in a power drill at a slow speed.

Mark the metal with a centre punch to locate the drill point, and clamp the work in a vice or to the bed of a vertical drill stand. Drill slowly and steadily, and keep the bit oiled. To drill a large hole, make a small pilot hole first to guide the larger drill bit.

When drilling sheet metal, the bit can jam and produce a ragged hole as it exits on the far side of the workpiece. As a precaution, clamp the work between pieces of plywood and drill through all three layers.

Masonry core drills
These are heavy-duty versions of the wood-working hole saw. Masonry core drills cut holes up to 150mm (6in) diameter in brick or stone walls for running new waste pipes to the outside.

Hole punch

Use a hole punch to make large holes in sheet metal. Having first marked out the circumference of the hole on the metal with spring dividers, lay the work on a piece of scrap softwood or plywood. Place the punch on the marked circle and tap it with a hammer, then check the alignment of the punched ring with the scribed circle. Reposition the punch and, with one sharp hammer blow, cut through the metal. If the wood crushes and the metal is slightly distorted, tap it flat again with the hammer.

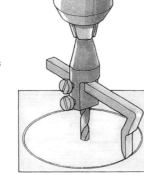

Tank cutter

Use a tank cutter to make holes for pipework in plastic or metal cold-water storage tanks.

● **Essential tools**
High-speed twist drills
Power drill
Bending springs
Soft mallet
Soldering iron
Gas torch

METAL BENDERS

Thick or hard metal must be heated before it can be bent successfully, but soft copper piping and sheet metal can be bent while cold.

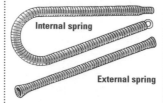

Internal spring

External spring

Bending springs

You can bend small-diameter pipes over your knee, but their walls must be supported with a coiled spring to prevent buckling.

Push an internal spring inside the pipe, or slide an external one over it. Either type of spring must fit the pipe exactly.

CURVED FORMERS

STRAIGHT FORMERS

Tube bender

With a tube bender, a pipe is bent over one of two fixed curved formers that are designed to give the optimum radii for plumbing and support the walls of the pipe during bending. Each has a matching straight former, which is placed between the pipe and a steel roller on a movable lever. Operating this lever bends the pipe over the curved former.

Soft mallet

Soft mallets have a head made of coiled rawhide, hard rubber or plastic. They are used in bending strip or sheet metal, which would be damaged by a metal hammer.

To bend sheet metal at a right angle, clamp it between stout battens along the bending line. Start at one end and bend the metal over one of the battens by tapping it with the mallet. Don't attempt the full bend at once, but work along the sheet, increasing the angle gradually and keeping it constant along the length until the metal lies flat on the batten. Tap out any kinks.

PIPE-FREEZING EQUIPMENT

To work on plumbing without having to drain the system, you can form temporary ice plugs in the pipework. The water has to be cold and not flowing.

Using freezing equipment

You can buy a kit containing an aerosol of liquid freezing gas, plus two plastic-foam 'jackets' to wrap round the pipework at the points where you want the water to freeze. Pierce a small hole through the wall of each jacket and bind it securely to the pipe **(1)**; then insert the extension tube through the hole **(2)** and inject the recommended amount of gas. It takes about five minutes for the ice plug to form in a metal pipe, and up to 15 minutes in a plastic one. If the job takes more than half an hour to complete, you will need to inject more gas.

Alternatively, hire jackets with cylinders of carbon dioxide; or an electric freezer connected to two blocks that you clamp over the pipework. An electric freezer will keep the water frozen until you finish the job and switch off.

1 Wrap a jacket around the pipe

2 Inject freezing gas inside the jacket

TOOLS FOR JOINING METAL

You can make permanent water-tight joints with solder, a molten alloy that acts like a glue when it cools and solidifies.

Mechanical fixings such as compression joints, rivets, and nuts and bolts are also used for joining metal.

SOLDERS

Solders are designed to melt at relatively low temperatures, but they will not work in the presence of water. When working on hot-water and cold-water plumbing, use a lead-free solder. It has a slightly higher melting point than the old lead solder and makes stronger joints.

FLUX

To be soldered successfully, a joint must be perfectly clean and free of oxides. Even after the metal has been cleaned with wire wool or emery, oxides form immediately, making a positive bond between the solder and metal impossible. Flux is therefore used to form a chemical barrier against oxidation.

Corrosive or 'active' flux, applied with a brush, dissolves oxides but must be washed from the surface with water as soon as the solder solidifies, or it will go on corroding the metal.

A 'passive' flux, in paste form, is used where it is impossible to wash the joint thoroughly. Although it does not dissolve oxides, it excludes them adequately for soldering copper plumbing joints and electrical connections.

Another alternative is to use wire solder containing flux in a hollow core. The flux flows just before the solder melts.

To flush flux from a central-heating system, fill it with water and let it heat up, then switch off and drain the system. This should be repeated a couple of times.

Soldering irons

For successful soldering, the work has to become hot enough for the solder to melt and flow – otherwise it solidifies before it can completely penetrate the joint. A soldering iron is used to apply the necessary heat.

Pencil-point iron

Tapered-tip iron

Use a low-powered pencil-point iron for soldering electrical connections. To bring sheet metal up to working temperature, use a larger iron with a tapered tip.

Tinning a soldering iron
The tip of a soldering iron has to be 'tinned' to keep it oxide-free. Clean the cool tip with a file; then heat it to working temperature, dip it in flux, and apply a stick of solder to coat it evenly.

 SEE ALSO: Soldering pipes 21, Bending pipes 23, Storage tanks 49, Spring dividers 74

Plumbing tools

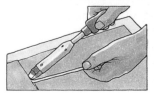

Using a soldering iron
Clean the mating surfaces of the joint to a bright finish and coat them with flux, then clamp the joint tightly between two wooden battens. Apply the hot iron along the joint to heat the metal thoroughly; and then run its tip along the edge of the joint, following closely with a stick of solder. The solder flows immediately into a properly heated joint.

Gas torch
Even a large soldering iron can't heat thick metal fast enough to compensate for heat loss from the joint, and this is very much the situation when you solder pipework. Although the copper unions have very thin walls, the pipe on each side dissipates so much heat that a soldering iron cannot get the joint itself hot enough to form a watertight soldered seal. You therefore need to use a gas torch with an intensely hot flame to heat the work quickly. The torch runs on liquid gas contained under pressure in a disposable metal canister that screws onto the gas inlet. Open the control valve and light the gas released from the nozzle, then adjust the valve until the flame roars and is bright blue. Use the hottest part of the flame – about the middle of its length – to heat the joint.

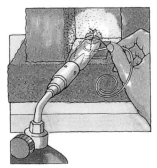

Hard soldering and brazing
Use a gas torch for brazing and hard soldering. Clean and flux the work – if possible with an active flux – then wire or clamp the parts together. Place the assembly on a fireproof mat or surround it with firebricks. Bring the joint to red heat with the torch, then dip a stick of the appropriate alloy in flux and apply it to the joint.
When the joint is cool, chip off hardened flux, wash the metal thoroughly in hot water, and finish the joint with a file.

Fireproof mat
Buy a fireproof mat from a plumber's merchant to protect flammable surfaces from the heat of a gas torch.

Hot-air gun
Some hot-air guns designed for stripping old paintwork can also be used for soft soldering. You can vary the temperature of an electronic gun from about 100 to 600°C. A heat shield on the nozzle reflects the heat back onto the work.

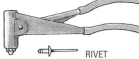
RIVET

Blind riveter
Join thin sheet metal with a blind riveter, a hand-operated tool with plier-like handles. It uses special rivets with long shanks that break off, leaving slightly raised heads on both sides of the work.

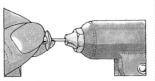
1 Insert the rivet

2 Squeeze the handles

Using a riveter
Clamp the two sheets together and drill holes right through the metal, matching the diameter of the rivets and spaced regularly along the joint. Open the handles of the riveter and insert the rivet shank in the head (**1**).

Push the rivet through a hole in the workpiece and, while pressing the tool hard against the metal, squeeze the handles to compress the rivet head on the far side (**2**). When the rivet is fully expanded, the shank will snap off in the tool.

SPANNERS AND WRENCHES

A professional plumber uses a great variety of spanners and wrenches on a wide range of fittings and fixings. However, there is no need to buy them all, since you can hire ones that you need only occasionally.

Open-ended spanner
A set of open-ended spanners is essential for a plumber or metalworker. Pipes generally run into a fitting or accessory, and the only tool you can use is a spanner with open jaws.
The spanners are usually double-ended (perhaps in a combination of metric and imperial sizes), and the sizes are duplicated within a set to enable you to manipulate two identical nuts simultaneously – on a compression joint, for example.

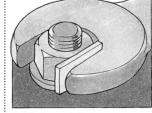

Achieving a tight fit
A spanner must be a good fit, or it will round the corners of the nut. You can pack out the jaws with a thin 'shim' of metal if a snug fit is otherwise not possible.

Ring spanner
Being a closed circle, the head of a ring spanner is stronger and fits better than that of an open-ended one. It is specially handy for loosening a corroded nut, provided you are able to slip the spanner over it.

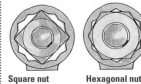

Square nut   **Hexagonal nut**

Choosing a ring spanner
Choose a 12-point spanner. It is fast to use and will fit both square and hexagonal nuts. You can buy combination spanners with a ring at one end and an open jaw at the other.

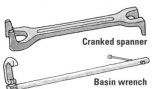

Box spanner
A box spanner is a steel tube with hexagonal ends. The turning force is applied with a tommy bar slipped through holes drilled in the tube. Don't use a very long bar: too much leverage may strip the thread of the fitting or distort the walls of the spanner.

Adjustable spanner
Having a movable jaw, an adjustable spanner is not as strong as an open-ended or ring spanner, but is often the only tool that will fit a large nut or one that's coated with paint. Make sure the spanner fits the nut snugly by rocking it slightly as you tighten the jaws; and grip the nut with the roots of the jaws. If you use just the tips, they can spring apart slightly under force and the spanner will slip.

Cranked spanner

Basin wrench

Cranked spanner and basin wrench
A cranked spanner is a special double-ended wrench for use on tap connectors.
A basin wrench (for the same job) has a pivoting jaw that can be set for either tightening or loosening a fitting.

Radiator spanner
Use this simple spanner, made from hexagonal-section steel rod, to remove radiator blanking plugs. One end is ground to fit plugs that have square sockets.

● **Essential tools**
Blind riveter
Set of open-ended spanners
Small and large adjustable spanners

☛ **SEE ALSO:** Tap connectors 24

Plumbing tools

Stillson wrench
The adjustable toothed jaws of a Stillson wrench are for gripping pipework. As force is applied, the jaws tighten on the work.

Chain wrench
A chain wrench does the same job as a Stillson wrench, but can be used on pipework and fittings with a very large diameter. Wrap the chain tightly round the work and engage it with the hook at the end of the wrench, then lever the handle towards the toothed jaw to apply turning force.

Smooth-jaw adjustable wrench
This older-style wrench is ideal for gripping and manipulating chromed fittings because its large smooth jaws will not damage the surface of the metal.

Strap wrench
With a strap wrench you can disconnect chromed pipework without damaging its surface. Wrap the smooth leather or canvas strap round the pipe, pass its end through the slot in the head of the tool, and pull it tight. Levering on the handle rotates the pipe.

Plier wrench
A plier wrench locks onto the work. It grips round stock or damaged nuts, and is often used as a small cramp.

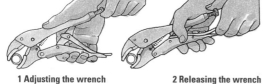

1 Adjusting the wrench   **2 Releasing the wrench**

Using a plier wrench
To close the jaws, squeeze the handles while slowly turning the adjusting screw clockwise (1). Eventually the jaws will snap together, gripping the work securely. To release the tool's grip on the work, pull the release lever (2).

● **Essential tools**
Plier wrench
Second-cut and smooth flat files
Second-cut and smooth half-round files

FILES
Files are used for shaping and smoothing metal components and removing sharp edges.

CLASSIFYING FILES

The working faces of a file are composed of parallel ridges, or teeth, set at about 70 degrees to its edges. A file is classified according to the size and spacing of its teeth and whether it has one or two sets of teeth.

Single-cut file

Double-cut file

A **single-cut file** has one set of teeth virtually covering each of its faces. A **double-cut file** has a second set of identical teeth crossing the first at a 45-degree angle. Some files are single-cut on one side and double-cut on the other.
    The spacing of teeth relates directly to their size: the finer the teeth, the more closely packed they are. Degrees of coarseness are expressed as number of teeth per 25mm (1in). Use progressively finer files to remove marks left by coarser ones.

● **File classification:**

● **Bastard file** – Coarse grade (26 teeth per 25mm), used for initial shaping.
● **Second-cut file** – Medium grade (36 teeth per 25mm), used for preliminary smoothing.
● **Smooth file** – Fine grade (47 teeth per 25mm), used for final smoothing.

CLEANING A FILE

● Soft metal tends to clog file teeth. When a file stops cutting efficiently, brush along the teeth with a fine wire brush, then rub chalk on the file to help reduce clogging in future.

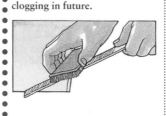

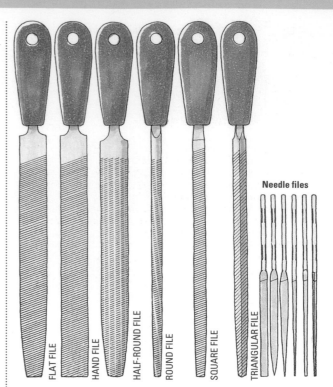

FLAT FILE   HAND FILE   HALF-ROUND FILE   ROUND FILE   SQUARE FILE   TRIANGULAR FILE

Needle files

Flat file
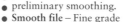
A flat file tapers from its pointed tang to its tip, in both width and thickness. Both faces and both edges are toothed.

Hand file
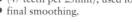
Hand files are parallel-sided but tapered in their thickness. Most of them have one smooth edge for filing up to a corner without damaging it.

Half-round file
This tool has one rounded face for shaping inside curves.

Round file
A round file is for shaping tight curves and enlarging holes.

Square file
Square files are used for cutting narrow slots and smoothing the edges of small rectangular holes.

Triangular file
Triangular files are designed for accurately shaping and smoothing undercut apertures of less than 90 degrees.

Needle files
These are miniature versions of standard files and are all made in extra-fine grades. Needle files are used for precise work and to sharpen brace bits.

FILE SAFETY

● Always fit a wooden or plastic handle on the tang of a file before you use it.

1 Fitting a file handle

2 Knock a handle from the tang

● If an unprotected file catches on the work, then the tang could be driven into the palm of your hand. Having fitted a handle, tap its end on a bench to tighten its grip (1).
    To remove a handle, hold the blade of the file in one hand and strike the ferrule away from you with a block of wood (2).

Plumbing tools

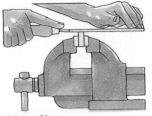

Using a file
When using any file, keep it flat on the work and avoid rocking it during forward strokes. Hold it steady, with the fingers of one hand resting on its tip, and make slow firm strokes with the full length of the file.

To avoid vibration, hold the work low in the jaws of a vice or clamp it between two battens.

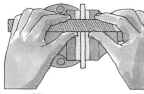

Draw filing
You can give metal a smooth finish by draw filing. With both hands, hold a smooth file at right angles to the work and slide the tool backwards and forwards along the surface. Finally, polish the workpiece with emery cloth wrapped round the file.

PLIERS

Pliers are for improving your grip on small components and for bending and shaping metal rod and wire.

Engineer's pliers
For general-purpose work, buy a sturdy pair of engineer's pliers. The toothed jaws have a curved section for gripping round stock and also have side cutters for cropping wire.

Slip-joint or waterpump pliers
The special feature of slip-joint pliers is a movable pivot for enlarging the jaw spacing. The extra-long handles give a good grip on pipes and other fittings. Use smooth-jaw pliers to grip chromed fittings.

FINISHING METAL

Before painting or soldering metal, always make sure it is clean and rust-free.

Wire brush
Use a steel-wire hand brush to clean rusty or corroded metal.

Wire wool
Wire wool is a mass of very thin steel filaments. It is used to remove file marks and to clean oxides and dirt from metals.

Emery cloth and paper
Emery is a natural black grit which, when backed with paper or cloth, is ideal for polishing metals. There is a range of grades from coarse to fine. For the best finish, use progressively finer abrasives as the work proceeds.

1 Glue paper to a board

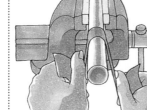

2 Clean a pipe with an emery strip

Using emery cloth and paper
To avoid rounding the crisp edges of a flat component, glue a sheet of emery paper to a board and rub the metal on the abrasive **(1)**.

To finish round stock or pipes, loop a strip of emery cloth over the work and pull alternately on each end **(2)**.

Buffing mop
Metals can be brought to a shine by hand, using a liquid metal polish and a soft cloth; but for a really high gloss, use a buffing mop in a bench-mounted power drill or grinder.

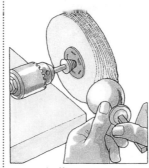

Using a buffing mop
After applying a stick of buffing compound (a fine abrasive with wax) to the revolving mop, move the work from side to side against the lower half, keeping any edges facing downwards.

WOODWORKING TOOLS

A plumber needs a set of basic woodworking tools in order to lift floorboards, notch joists for pipe runs, and attach pipe clips.

Reseating tool
If the seat of a tap has become so worn that even fitting a new washer won't produce a perfect seal, use a reseating tool to grind the seat flat.

Remove the tap's headgear and jumper, then screw the cone of the reseating tool into the body of the tap. Turn the knurled adjuster to lower the cutter onto the worn seat, and then turn the tommy bar to regrind the metal.

● **Essential tools and materials**
Engineer's pliers
Wire brush
Wire wool
Emery cloth and emery paper

MAXIMUM LENGTHS FOR DOMESTIC CIRCUITS

TYPE OF CIRCUIT	Max. floor area in sq m	Cable size in mm²	Size of earth wire in mm²	Current rating of circuit fuse	Max. cable length using cartridge fuse	Current rating of MCB	Max. cable length using MCB
				USING FUSES		USING MCBs	
RING CIRCUIT	100	2.5	1.5	30amp	68m	32amp	68m
RADIAL CIRCUIT	20	2.5	1.5	20amp	37m	20amp	34m
	50	4	1.5	30amp	19m	32amp	21m
COOKER up to 13.5kW		4	1.5	30amp	19m	32amp	21m
COOKER from 13.5 to 18kW		6	2.5			40amp	27m
IMMERSION HEATER up to 3kW		2.5	1.5	15amp	39m	16amp	39m
SHOWER up to 10.3kW		10	4	45amp	46m	45amp	46m
SHOWER from 10.3 to 10.8kW		10	4			50amp	44m
STORAGE HEATER up to 3.375kW		2.5	1.5	15amp	34m	16amp	34m
STORAGE FAN HEATER up to 6kW		4	1.5	30amp	32m	32amp	32m
FIXED LIGHTING excluding switch drops		1	1	5amp	83m	6amp	83m
		1.5	1	5amp	126m	6amp	126m

☞ **SEE ALSO:** Replacing washers 10, Dismantling taps 33

Artificial ventilation

Severe electric shock can make a person stop breathing. Once you have freed them from the electricity supply (without grasping the victim's body directly – see right), revive them by means of artificial ventilation.

Clear the airway

First, clear the victim's airway. To do this, loosen the clothing round the neck, chest and waist, make sure that the mouth is free of food, and remove loose dentures.

Clear the mouth of food or loose dentures.

Lay the person on his or her back and carefully tilt the head back by raising the chin. This prevents the victim's tongue blocking the airway and may in itself be enough to restart the person's breathing. If it doesn't succeed in doing so quickly, try more direct methods of artificial ventilation.

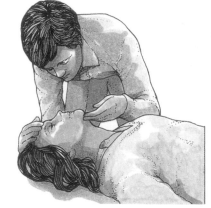

Tip the head back to open the airway.

Mouth-to-mouth

Keeping the victim's nostrils closed by pinching them between thumb and forefinger, cover the mouth with your own, making a seal all around **(1)**. Blow firmly and look for signs of the chest rising. Remove your lips and allow the chest to fall. Repeat this procedure, breathing rhythmically into the mouth every six seconds. After ten breaths, phone the emergency services. Then continue with the artificial ventilation till normal breathing resumes or expert help arrives to take over.

Mouth-to-nose

If injuries to the face make mouth-to-mouth ventilation impossible, follow a similar procedure but keep the victim's mouth covered with one hand and blow firmly into the nose **(2)**.

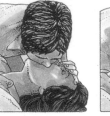

1 Mouth-to-mouth ventilation　**2 The mouth-to-nose procedure**

Reviving a baby

If the victim is a baby or small child, cover both the nose and the mouth at the same time with your own mouth and proceed as for mouth-to-mouth ventilation (see below, left), but breathe every three seconds.

To give artificial ventilation to a small child, cover the nose and mouth.

Recovery position

Once breathing has started again, put the victim in the recovery position. Turn him or her face down with the head turned sideways and tilted up slightly. This keeps the airway open and will also prevent vomit being inhaled if the person is sick. Lift one leg out from the body and support the head by placing the person's left hand, palm down, under his or her cheek. Keep the casualty warm with blankets until help arrives.

If someone receives an electric shock and is still in contact with its source, turn off the current either by pulling out the plug or by switching off at the socket or consumer unit. If this is not possible, don't take hold of the person as the current may pass through you too. Pull the victim free with a scarf or dry towel, or knock their hand free with a piece of wood. As a last resort, free the victim by taking hold of their loose clothing – but without touching the body. Don't attempt to move anyone who has fallen as a result of electric shock – except to place them in the recovery position. Wrap them in a blanket or coat to keep them warm until they can move. Once the person can move, treat their electrical burns by reducing the heat of the injury under slowly running cold water. Then apply a dry dressing and seek medical advice.

Isolating the victim
If a person sustains an electric shock, turn off the supply of electricity immediately, either at the consumer unit or at a socket **(1)**. If this is not possible, pull the victim free with a dry towel, or knock their hand free of the electrical equipment **(2)** with a piece of wood or a broom.

Place the victim on his or her side with the head turned sideways and one leg out from the body.

☞ SEE ALSO: Bathroom safety 69

Glossary

Adaptor
A device that is used to connect more than one appliance to a socket outlet.

Airlock
A blockage in a pipe caused by a trapped bubble of air.

Appliance
A functional piece of equipment connected to the plumbing – a basin, sink, bath etc.

Back-siphonage
The siphoning of part of a plumbing system caused by the failure of mains pressure.

Balanced flue
A ducting system which allows a heating appliance, such as a boiler, to draw fresh air from, and discharge gases to, the outside of a building.

Bore
Hollow part of a pipe or tube.

Burr
Rough raised edge left on a metal workpiece after cutting or filing.

Cap-nut
The nut used to tighten a fitting onto pipework.

Cesspool
A covered or buried tank for the collection and storage of sewage.

Chase
The groove cut in masonry to accept a pipe or cable. *Or* To cut such grooves.

Circuit breaker
A special switch installed in a consumer unit to protect an individual circuit. Should a fault occur, the circuit breaker will switch off automatically.

Consumer unit
A box, situated near the meter, which contains the fuses of MCBs protecting all the circuits. It also houses the main isolating switch that cuts the power to the whole building.

Cistern
A water-storage tank such as found in the roof of a house.

Draincock
Tap from which a plumbing system or single appliance is drained.

Economy 7
An Electricity Company scheme which allows you to charge storage heaters and heat water at less than half the general-purpose rate.

Float valve
A water inlet which is closed by the action of a float-operated arm when the water in a cistern reaches the required level.

Earth
A connection between an electrical circuit and the earth (ground).

Fuse
A protective device containing a thin wire that is designed to melt at a given temperature caused by an excess flow of current on a circuit.

Gully
The open end of a drainage system at ground level, containing a water-filled trap.

Head
The height of the surface of water above a specific point – used as a measurement of pressure; for example, a head of 2m.

Hopper head
The funnel-shaped end of a drainage pipe that receives the discharge from other wastepipes.

Immersion heater
An electrical element designed to heat water in a storage cylinder.

Overflow pipe
A drainage pipe designed to discharge water which has risen above its intended level within a cistern.

PTFE
Polytetrafluorethylene – used to make tape for sealing threaded plumbing fittings.

Rising main
The pipe which supplies water under mains pressure, usually to a storage cistern in the roof.

Septic tank
A sewage-storage tank, similar to a cesspool, but the waste is treated to render it harmless before it is discharged underground or into a local waterway.

Shoe
The component forming the lower end of a vertical drainage pipe and which throws water clear of the wall into an open gully.

Stopcock
Valve which closes a pipe to prevent the passage of water.

Storage heater
A space-heating device that stores heat generated by cheap night-rate electricity, then releases it during the following day.

Supplementary bonding
The connecting to earth of exposed metal appliances and pipework within a bathroom or kitchen.

Thermostat
A device which maintains a heating system at a constant temperature.

Trap
A bent section of pipework, containing standing water to prevent the passage of noxious sewer gases.

Water closet – WC
A lavatory flushed by water.

Water hammer
Vibration caused by fluctuating water pressure within a plumbing system.

Wiring Regulations
A code of professional practice laid down by the Institution of Electrical Enginners.

If you have enjoyed this book, why not build on your expertise with other Collins titles?

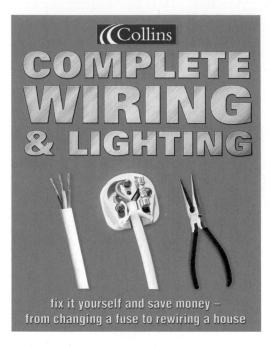

A completely revised and updated edition
of the number one guide to home electrics

72pp   £8.99

PB   ISBN 0 00 716440 8

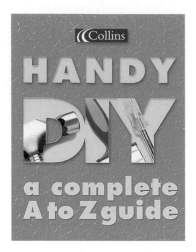

A pocket-sized A-Z guide for using
on the job

448pp   £8.99

PB   ISBN 0 00 714668 X

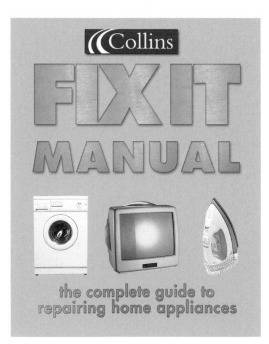

Easy-to-follow solutions for fixing and
maintaining everyday electrical appliances

304pp   £25.00

HB   ISBN 0 00 412993 8

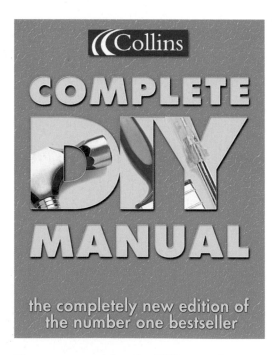

The most comprehensive DIY manual ever
published – over 1 million copies sold

552pp   £24.99

HB   ISBN 0 00 414101 6

To order any of these titles please telephone **0870 787 1732**

For further information about Collins books visit our Website: **www.collins.co.uk**